RENEWALS 458-4574

Fluidized Bed Technology

Principles and Applications

Fluidized Bed Technology
Principles and Applications

J R Howard

Adam Hilger, Bristol and New York

© J R Howard 1989

All rights reserved. No part of this publication may be reproduced, stored in a retrieval system or transmitted in any form or by any means, electronic, mechanical, photocopying, recording or otherwise, without the prior permission of the publisher. Multiple copying is only permitted under the terms of the agreement between the Committee of Vice-Chancellors and Principals and the Copyright Licensing Agency.

British Library Cataloguing in Publication Data

Howard, J. R.
  Fluidized bed technology
  1. Gas fluidized beds
  I. Title
  660.2′84292

  ISBN 0-85274-055-7

Library of Congress Cataloging-in-Publication Data

Howard, J. R.
  Fluidized bed technology: principles and applications/J. R. Howard.
    p. cm.
  Includes bibliographies and index.
  ISBN 0-85274-055-7
  1. Fluidization. I. Title.
  TP156.F65H68 1989
  660.2′84292–dc19

The author and IOP Publishing Ltd have attempted to trace the copyright holder of all the figures reproduced in this publication and apologize to copyright holders if permission to publish in this form has not been obtained.

Consultant Editor: **Mr A E De Barr**

Published under the Adam Hilger imprint by IOP Publishing Ltd
Techno House, Redcliffe Way, Bristol BS1 6NX, England
335 East 45th Street, New York, NY 10017-3483, USA

Typeset by KEYTEC, Bridport, Dorset
Printed in Great Britain by J W Arrowsmith Ltd, Bristol

Contents

Preface		ix
Acknowledgments		xi
1	**Introduction**	**1**
	1.1 Processes involving contact between solid particles and a fluid	1
	1.2 Packed (or fixed) beds	4
	1.3 Fluidized beds	5
	1.4 Treatment of the subject in this book	13
	References	13
	Bibliography	14
2	**Particles and fluidization**	**15**
	2.1 Physical properties of solid particles	15
	2.2 Classification of particles according to fluidization characteristics	24
	2.3 Pressure drop across packed beds	26
	2.4 Minimum fluidizing velocity and its determination	30
	2.5 Two-phase theory of fluidization bubbles and fluidization regimes	38
	2.6 Mixing, elutriation and transport of solids	47
	Examples for chapters 1 and 2	63
	References	66
	Bibliography	69
3	**Fluidized bed heat transfer**	**70**
	3.1 Modes of heat transfer	70
	3.2 Heat transfer in beds of particles	72
	3.3 Estimation of bed-to-surface heat transfer coefficients	84
	3.4 Heat transfer between the bed, distributor, containing walls, immersed tubes or components	89
	3.5 Heat transfer to surfaces located above the bed free surface	100

vi Contents

    3.6 Concluding remarks   101
          Examples   101
          References   104
          Bibliography   105

4 Design of simple fluidized beds   107
    4.1 Introduction   107
    4.2 Estimation of bed dimensions and fluidizing velocity   109
    4.3 Transport disengaging height   115
    4.4 Distributors   116
    4.5 Heat removal from fluidized beds   120
    4.6 Optimum size of a fluidized bed reactor   137
    4.7 Concluding remarks   141
          References   142
          Bibliography   142

5 Fluidized bed combustion   143
    5.1 Introduction   143
    5.2 Combustion systems for solid fuels generally   143
    5.3 Fluidized bed combustion of solid fuels   150
    5.4 Size of fluidized bed combustion systems   154
    5.5 Efficiency of fluidized bed combustion equipment   163
    5.6 Combustion of fuel particles in a fluidized bed   166
    5.7 Distinction between boilers and furnaces   179
    5.8 Methods of starting up   179
    5.9 Circulating or 'fast' fluidized bed combustion systems   181
    5.10 Control of emissions   182
    5.11 Concluding remarks   186
          Examples   186
          References   188
          Bibliography   191

6 Closure   193
    6.1 Introduction   193
    6.2 Gasifiers   193
    6.3 Dryers   194
    6.4 Metallurgical heat treatment furnaces   197
    6.5 Solids transport systems   198
    6.6 Flue gas desulphurization   199
    6.7 Fluidized bed catalytic cracking   200
    6.8 Design practice and scale-up   202
    6.9 Final comments   204
          References   204

Appendix A	207
Index	209

Preface

The primary objective of this book is to help beginners and students gain sufficient insight into the subject of fluidized bed technology, from which they can form their own appreciation of the capabilities and limitations of the technology and proceed, if necessary, to greater expertise. The fluid used for fluidization of a bed of solid particles can be either a liquid or a gas, but this book concentrates entirely upon beds of particles fluidized by gases, for reasons of space.

Accordingly, the early part of the book begins with the fundamental subject matter, describing how a bed of solid particles may behave when a gas is passed upwards through the bed at a sufficiently high rate of flow for the bed to assume fluid-like properties. The book then develops, sometimes using worked examples, to illustrate the considerations involved in the design of equipment based on fluidized bed technology.

The treatment is pitched at a modest level such that anyone with a basic understanding of engineering and science should be able to comprehend. It is also hoped that the book will be valuable to experienced engineers, particularly if they are having to train staff, teachers in engineering departments of colleges, universities and polytechnics and also managers or consultants who may need to gain a rapid appreciation of the technology.

The quantity of literature devoted to fluidized beds, especially research and development papers, is vast and the degree of expertise and knowledge required to design, construct and operate modern fluidized bed systems successfully requires a large investment of time and effort. Experience has shown that anyone without prior knowledge, who desires to gain an appreciation of any particular technology, is likely to have difficulty in knowing where to start. It is hoped that this book will help readers to make such a start with fluidized bed technology. In addition to the instructional aspects of this book, a selection of literature references has been included from which the reader may obtain further guidance.

I am extremely grateful to many friends and colleagues with whom I have discussed various topics over the years and from whom I have received invaluable criticism and advice as the work of writing progressed. My thanks are especially due to Dr John Botterill, Mr Derek Hickson, Dr Bernie Gibbs, Mr David Reay and the referees appointed by the publishers. Finally, I shall always owe a debt of gratitude to my former colleague, the late Professor Douglas Elliott, who first stimulated my interest in fluidized beds and who has been justly described† as one of the real heroes of the story.

J R Howard

† Ehrlich S 1976 in *Proc. 4th Int. Conf. Fluidized Bed Combustion, 1975* (McLean, Virginia: MITRE Corporation) pp 15–20.

Acknowledgments

I am grateful to the following for making various diagrams available to me and for giving permission to publish them in this work.

American Institute of Chemical Engineers for figure 1.8.
Elsevier Sequoia S A for figure 2.5.
Elsevier Applied Science Publishers Ltd for figures 2.5, 5.7, 5.8, 5.11.
Academic Press for figure 4.3.
Professor Prabir Basu for figures 5.16 and 5.17.
J S Harrison pp British Coal for figure 6.1.
Wellman Process Engineering Ltd and John Brown Engineers and Constructors Ltd for figure 6.2.
John Wiley & Sons Ltd and Dr D Reay, of Engineering Sciences Division, AERE, Harwell for figures 6.3 and 6.5.
The Wolfson Heat Treatment Centre for figure 6.4.
Pergamon Press PLC for figure 6.6.

1 Introduction

1.1 Processes Involving Contact Between Solid Particles and a Fluid

A great many industrial processes involve contact and interaction between solids and fluids (i.e. gases or liquids). Examples include combustion, gasification of solid fuels, shales or solid wastes, drying of particles, calcining, particle heating, regenerative heat exchangers, oxidation or reduction of ores, metal surface treatments and catalytic and thermal cracking.

It may be helpful to distinguish roughly between the functions of fluids and solids and give examples of practical plant. This is done in table 1.1.

Table 1.1 The functions of fluids and solids and some examples of associated plants.

Fluid function	Solid function	Examples of plant
Heat carrier	Non-reactive feedstock	Solid heaters, regenerative heat exchangers, foundry sand reclaimers, dryers
Mass carrier	Non-reactive feedstock	Pneumatic or hydraulic solids transporters
Heat and mass carrier	Non-reactive feedstock	Dryers
Chemical reactant	Chemical reactant	Combustors, gasifiers, catalytic cracking, catalyst regeneration, ore reduction or oxidation plant
	Active particles for chemical reaction, inert particles as heat transfer	Combustors, metallurgical surface treatment and heat treatment furnaces
Agitation only	Non-reactive	Mixers, classifiers

2   Introduction

It will be seen from table 1.1 that several combinations of the functions of solids and fluids can arise in practical plant. Thus, when considering processes or plants it is necessary to be clear as to the particular purpose served by the fluids and the solids. Heating and drying of solids, for example, involve heat and mass transfer only, whereas combustors, gasifiers etc have the additional complication of chemical reactions which have to be carried out simultaneously with heat and mass transfer. However, with both categories it is sufficient at this stage to say that, broadly speaking, the principles used to bring fluids and solids into contact are similar.

Conceptually, the easiest way to perform such processes is to arrange for the fluid to flow through a bed of the solid particles, percolating through the interstices between the particles, as shown in figure 1.1. The simple 'packed bed' of particles shown in figure 1.1 comprises a containing vessel with a porous base through which fluid can flow upwards or downwards among the particles. (Frequently, the fluid is made to flow downwards rather than upwards because when flowing upwards the fluid velocity is limited to that which is the smaller of (i) that for incipient fluidization of the bed (see §1.3.2) or (ii) that resulting in an excessive amount of the finer particles being blown out of the system. With downward flowing fluid, the limit to fluid velocity is set by the allowable functional pressure drop across the bed.) In either case the fluid makes contact with the surface of the particles and continuous fluid flow ensures exposure of the particle surface to a fresh fluid continuously. The nature of the fluid flow (e.g. laminar, turbulent, transitional) at the particle/fluid interface influences the *rate* of interaction between fluid and particle, for example the heat or mass transfer rate or the chemical reaction rate at the interface.

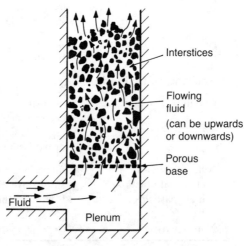

Figure 1.1 A packed bed.

Other ways of bringing a fluid into contact with the surface of particles include the dropping of particles down a vertical duct containing the fluid, as shown in figure 1.2; mechanical stirring or agitation (see figure 1.3); the use of a 'spouted bed', as shown in figure 1.4, in which a jet of fluid at high velocity pierces through the bed, entraining and agitating the particles; or the use of a moving bed (see figure 1.5). Each of these ways of bringing fluid into contact with the surface of particles has its merits and its deficiencies. Deciding which of them is the most

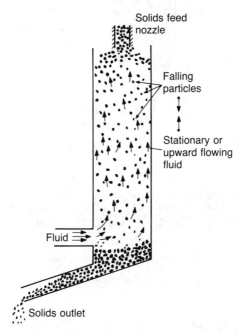

Figure 1.2 A 'falling cloud' duct.

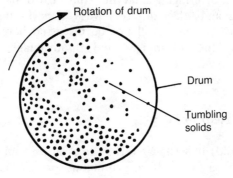

Figure 1.3 The agitation of solids by a rotating drum.

4 Introduction

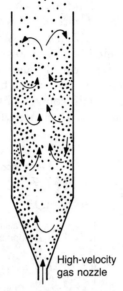

Figure 1.4 A spouted bed.

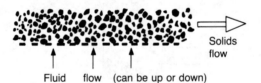

Figure 1.5 A flowing packed bed.

suitable for a prescribed task is not always easy and depends upon the constraints (including economics and convenience) within which the process has to be accomplished. Indeed, two different designers faced with the same task of bringing fluid into contact with particles may have differing opinions and select different ways of performing the process. The procedure for selection will not be discussed here, however, but table 1.2 outlines the advantages and disadvantages of alternative methods.

1.2 Packed (or Fixed) Beds

A packed (or fixed) bed (figure 1.1) suffices in principle for processing solids a batch at a time. If a continuous process is required, the unprocessed solids must flow into the container and the processed solids be removed at the desired rate. Such a system becomes a flowing packed bed. Static packed beds are unsuitable for continuous processes because of the difficulty of transporting solids into and out of the system easily.

There are other difficulties with packed beds such as the existence of temperature gradients within the bed if the fluid/solid or gas/solid reaction is exothermic; this can result in sintering of the particles. Heat transfer to and from the bed is relatively poor, requiring a large surface area to effect it. Further, if the fluid velocity has to be high (to satisfy the requirement of high throughput), then only large particles can be processed because the smaller ones may be transported out of the bed by entrainment in the gas. On the other hand, the pressure drop across the bed is relatively small.

Fluid flowing through the spaces between the particles (interstices) exerts a drag force on the particles and this force may be large enough to disturb the arrangement of the particles within the bed. If the upward velocity of the fluid through the bed is raised progressively, a situation will eventually arise where the fluid drag exerted on the bed of particles is sufficient to support its entire weight. The bed is then said to be 'incipiently fluidized' and it exhibits fluid-like properties. The bed will flow under a hydrostatic head, the free surface will remain horizontal if the containment is tilted and low-density objects will float (see figure 1.6(*a*), (*b*), (*c*) for illustration). This leads naturally to the exploitation of the phenomenon of fluidization to, for example, the transport of solids and to the examination of what happens when fluid flow is increased beyond that required for incipient fluidization.

1.3 Fluidized Beds

1.3.1 Components

From the discussion in §1.2 it will be seen that the most easily perceived parameter dictating whether a bed is fluidized or packed is the velocity at which the fluid passes upwards through an unrestrained bed of particles. Both types of bed require a containing vessel with a porous base through which the fluid can be introduced to the bed. The exact form of the porous base is a matter of design choice, ranging from a plate with a large number of small holes drilled in it to a block of sintered ceramic, powder metal, stand pipes or bubble caps. Whatever the construction, the most important function of the porous base is that it distributes the fluid across the base of the bed uniformly. The porous base is therefore almost universally called a 'distributor'.

Table 1.2 A comparison of alternative gas-solid contacting methods.

Method	Advantages	Disadvantages
Packed beds (figure 1.1)	High conversion provided that control of gas distribution and temperature is good (not easy) Long gas residence time High flow rates through constrained bed at expense of higher pressure drop	Low gas velocity—large reactor size Only large particles Steep temperature gradients and large heat transfer areas required with exothermic reactions Danger of particle sintering and blockage of reactor Suitable only for batch processes Non-uniformity of product
Fluidized beds (figure 1.7) bubbling regime	Good gas–solid contact Good particle mixing Uniform temperature and control of process giving uniform quality of products High bed-to-surface heat transfer coefficients Can use wider particle size range Ease of transport of solids into and out of reactor	Lower conversion Pressure drop and necessary pumping power increase with deep beds Erosion of vessel and pipes and production of fines by attrition Elutriation of fines can limit performance Gas by-pass can be excessive Size range of particle to be used is limited Segregation of particles
Falling cloud (figure 1.2)	Simplicity of construction Low pumping power Can operate with dirty gases	Dependent upon reliable means of dispersing particles across duct section Non-uniform particle residence time across duct section Gas velocity distribution non-uniform Close particle size range required Large vessel per unit mass flow of solids or per unit heat rate

Table 1.2 (*cont*)

Method	Advantages	Disadvantages
Flowing packed bed (figure 1.5)	Small cross sectional area of duct for solids transport Uniform temperature distribution Relatively little elutriation Large thermal capacity for transporting heat High conversion	Particle segregation Maximum gas velocity limited to minimum fluidizing velocity of solids Heat transfer coefficients low Close particle size range required
Spouted bed (figure 1.4)	Good agitation of particles which are too coarse or non-uniform for good fluidization Interparticle collisions inhibit agglomeration of particles or expose fresh surface of particles Regular cyclic motion of solids Spouts can be used to prevent local defluidization Spouted bed dryers—cheap—mechanically simple	High pressure drop, particularly on starting Particle attrition if particle residence times are long or particles are fragile Limited to relatively large particles Possibility of erosion

8   Introduction

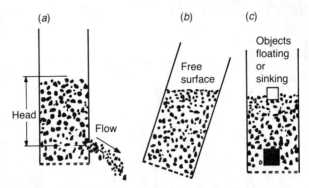

Figure 1.6 The behaviour of incipiently fluidized particles.

The containing vessel of a fluidized bed has to extend to a sufficient height above the free surface of the bed to allow space for particles which can be carried upwards from the free surface to disengage from the fluid stream and fall back into the bed. If such a 'particle disengagement height' is not provided, particles will be lost from the bed in the off-gas stream.

1.3.2 Fundamental fluidized bed behaviour

Beds of particles fluidized by a *gas behave differently from those fluidized by a liquid* once the gas flow rate is raised above that required to produce 'incipient fluidization'. This flow rate is normally quoted as a *velocity* and is termed the 'minimum fluidizing velocity', U_{mf}, defined simply as

$$U_{mf} = \frac{\dot{V}_{mf}}{A} \quad (1.1)$$

where $\dot{V}_{mf}$ is the volume flow rate at incipient fluidization and A is the cross sectional area of the bed containing vessel.

Example 1.1. If a bed of particles is contained in a vessel 500 mm in diameter and the volume flow rate of the fluidizing gas or liquid at incipient fluidization is 0.05 m³ s⁻¹, then the minimum fluidizing velocity U_{mf} is

$$U_{mf} = 0.05 \, \frac{m^3}{s} \, \frac{4}{\pi \times 0.5^2 \, m^2} = 0.255 \text{ m s}^{-1}. \quad (1.2)$$

Likewise, the fluidizing velocity U at any other condition is given by

$$U = \frac{\dot{V}}{A} \quad (1.3)$$

where $\dot{V}$ and A are the volume flow rate and cross sectional area of the containing vessel, respectively.

Note that the fluid velocity through the interstices between particles must be greater than the velocities calculated as above because the particles obstruct part of the cross sectional area of the containing vessel. However, because of the difficulty of determining the actual unobstructed area and its wide variation throughout the bed, the above definitions of fluidizing velocities are the simplest practical ones.

At fluid velocities up to incipient fluidization (see figure 1.7) the general behaviour of the bed is little affected whether the fluid is liquid or gas, but as the fluid velocity is increased further the behaviour when fluidized by liquid diverges from that when fluidized by gas. With liquids, uniform expansion of the bed occurs with increase in velocity until particles are carried out of the bed, whereas with gas fluidized systems, expansion is not uniform (except with fine powders) and instabilities develop in the form of cavities or bubbles. These are responsible for mixing of the solids. A body of knowledge of both liquid fluidized and gas fluidized systems has been built up but space is insufficient to write about both types.

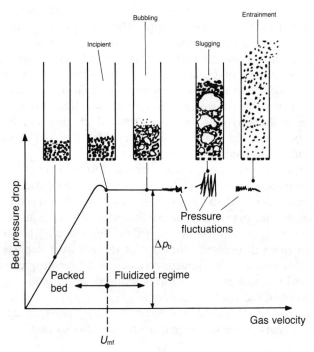

Figure 1.7 Bed behaviour with gas velocity changes.

Accordingly, *this book is devoted almost entirely to features of beds of particles which are fluidized by gases rather than by liquids* and we proceed now to consider the behaviour of gas fluidized systems in greater detail.

For simplicity at this stage, assume that the particles are of a size and individual density which is compatible with their being able to be fluidized satisfactorily; there are categories of particles which are difficult to fluidize and these are considered later in §2.2.2. Suppose the particles are contained in a vessel with a well designed distributor in the base and a gas is passed into the bed at a progressively increasing rate. During such an experiment, three observations need to be made, namely (i) a visual observation of the bed, (ii) the pressure drop across the bed itself and (iii) the depth of the bed.

Figure 1.7 shows the broad sequence of events as the gas flow increases progressively from zero. At first the appearance of the bed does not change and the pressure drop rises with flow rate, reaching a maximum value at the point of incipient fluidization. Increasing the velocity above the minimum fluidizing velocity U_{mf} does not result in an increase in bed pressure drop. At first the particles tend to rearrange themselves to provide more space around themselves so as to accommodate the extra gas flow, i.e. the bed voidage ε increases, so that the bed as a whole expands to a greater depth. The extent of this expansion and the increase in fluidizing velocity over which it is sustained depends upon the nature of the particles, as will be discussed later in §2.5.3. However, at some point bubbles or particle-free cavities are formed among the particles and these bubbles (non-uniformly distributed) rise through the bed, bursting when they reach the free surface, scattering particles into the above-bed region, from which they fall back to the bed. The bed then has two regions or 'phases': the 'bubble phase' and the 'particulate (or continuous) phase'.

A bubbling action causes the particles to mix continuously and thereby promote uniformity of the bed temperature or composition, but bubbling can also lead to an excessive by-pass of unreacted fluidizing gas; a compromise between these conflicting features has to be struck by the bed designer. At sufficiently high fluidizing velocity, however, some of the particles will become entrained in the gas as it departs from the bed. A further increase in fluidizing velocity entrains progressively more particles and the bed pressure drop is reduced until, finally, all particles are blown out of the containing vessel.

Once the bed is fluidized, the pressure drop across it, Δp_b, will be sufficient to support the full weight of the particles so that

$$\Delta p_b = \frac{M}{\rho_p A}(\rho_p - \rho_g)g \qquad (1.4)$$

where M is the mass of particles, ρ_p is the particle density, ρ_g is the fluidizing gas density, A is the cross sectional area of the bed containment and g is the gravitational acceleration.

Except for operations carried out at high static pressure, the density of the fluidizing gas is negligible compared with that of the particles, so that equation (1.4) may be simplified to

$$\Delta p_b = \frac{Mg}{A}. \tag{1.5}$$

Equations (1.4) and (1.5) imply that either there is no interaction between the bed of particles and its containing wall or that no energy is degraded in the bed; such degradation can arise due to collisions between a particle and a wall or between two particles. The effect of such interactions manifests itself as an increased pressure drop across the bed. Conditions under which such interactions are of sufficient magnitude to cause the pressure drop to be larger than that predicted by equations (1.4) and (1.5) arise, for example, when operating in the 'slugging' regime (see §2.5.4). The increase in pressure drop across the bed during the packed bed part of the pressure drop/fluidizing velocity curve (figure 1.7) depends upon how firmly the bed is packed before any fluid is passed through it. Firstly, at incipient fluidization, as shown in figure 1.7, a small 'hump' in the curve may be recorded, particularly during the very first attempt to fluidize the bed. This arises because extra pressure is required to 'unlock' the particles from their pattern of packing. Once they have been released from this pattern, the voidage, i.e. the fraction of the total volume of the bed which is occupied by the spaces between the particles, increases from its packed bed value, ε_{pb}, to a slightly larger value, ε_{mf}, at minimum fluidizing velocity; the pressure drop then falls back to a value sufficient to support the weight of particles in the bed. Secondly, one may also expect the slope of the pressure drop/fluidizing velocity graph in the packed bed region to be steeper when the bed is firmly compacted than when loosely packed, because of the smaller voidage.

1.3.3 'Fast' fluidization—circulating fluidized beds

For reasons of simplification, the above description of bed behaviour exemplified by figure 1.7 has deliberately left out any consideration of a fluidization regime which has assumed considerable importance and has grown in application over recent years, namely, the 'fast' fluidization regime; the corresponding fluidized beds are known as 'fast' or sometimes 'circulating' fluidized beds.

Consider now the fluidized bed system depicted by figure 1.8(*a*) in which a steady flow of particles for processing is supplied continuously to a fluidized bed reactor. The fluidizing velocity is such that most of the processed particles are withdrawn from the bed, the remainder being entrained in the exhaust gas. A cyclone in the exhaust gas line separates these entrained particles from the gas. These particles captured by the cyclone may be sufficiently processed to add to the main bulk of particles withdrawn from the bed, they may be rejected or they may be recycled to the bed for reprocessing according to the economics or other requirements of the plant. If the fluidizing velocity is low, the fraction of fine particles transported out of the bed is likely to be small, the main solids discharge from the plant will be from the bed, as shown in figure 1.8(a), and there the bed will have a reasonably clearly visible free surface. The mean residence time of the particles in the system is long since most of them reside in the bed until they overflow.

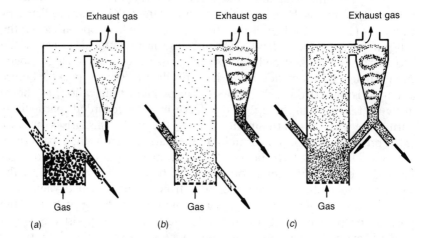

Figure 1.8 Various fluidization regimes for continuous operation. (a) Moderate fluidizing velocity. (b) Higher fluidizing velocity with high bed expansion and particle entrainment. (c) 'Fast' fluidization (high fluidizing velocity and high particle recycle rate). (Redrawn from Reh 1971.) Reproduced by permission of the American Institute of Chemical Engineers.

If, however, the fluidizing velocity is increased to a value sufficient to expand the bed to such an extent that most of the particles fed to the bed are captured by the cyclone, the free surface of the bed will not be clearly seen and the voidage of the gas/particle suspension will be large, but increasing with distance above the distributor, as shown in figure 1.8(b). Further, the mean residence time of the particles is short since most of them are swept through the system by the gas.

In between these two extremes is the 'fast' fluidization regime, depicted by the reactor system in figure 1.8(c), where the particles captured by the cyclone are recycled back to the reactor at a high rate, maybe at several times the input feed rate, increasing their mean residence time in the reactor and reducing the voidage of the expanded bed. The intense mixing leads to a uniformity of temperature and high rates of heat and mass transfer between gas and particle. Reh (1971) has shown that this fluidization regime lies in the transitional range between a fluidized bed and pneumatic transport. Several variants of this type of fluidized bed system have been developed; further details can be found from the references given at the end of this Chapter and from Chapter 5.

1.4 Treatment of the Subject in this Book

Fluidized bed technology has been utilized to carry out many diverse industrial processes involving gas–particle contacting. Examples range from metallurgical processes (one of the earliest patents (Robinson (1879) exploited fluidized bed technology for roasting of ores, although the term 'fluidized bed' was not mentioned) – carrying out many different kinds of chemical reactions (such as catalytic cracking) – burning coals or low-grade fuels – incineration – drying wet products – producing gas from coal – transporting particulate matter pneumatically – heat recovery and so on.

With the exception of Chapter 6, this book is intended to provide fundamental matter which, over decades of endeavour, has led to the development of modern industrial fluidized bed plant.

Accordingly, Chapter 2 considers properties of solid particles which are especially relevant to fluidized beds because of their important influence on bed behaviour; then further aspects of fluidization are considered. Chapter 3 describes mechanisms of heat transfer in fluidized beds and associated fundamental matter. Chapters 4 and 5 use the preceding subject matter to illustrate, partly with the aid of worked examples, the design of simple fluidized beds and fluidized bed combustion systems. The book concludes with Chapter 6 which gives brief notes about other applications and problems of scale-up of fluidized bed equipment. Throughout the work, selected references are given to enable the reader to pursue the subject further.

References

Reh L 1971 Fluidized bed processing *Chem. Eng. Prog.* **67** 58–63
Robinson C E 1879 Furnaces for roasting ores *US Patent Specification* 212 508

Bibliography

Basu P (ed.) 1986 *Circulating Fluidized Bed Technology* (Toronto: Pergamon)
Yerushalmi J 1986 in *Gas Fluidization Technology* ed. D Geldart (Chichester: Wiley – Interscience) ch 7

2 Particles and Fluidization

2.1 Physical Properties of Solid Particles

2.1.1 Particle size and shape

If a particle is spherical in shape, then the size of the particle is clearly characterized by its diameter. However, spherical particles are seldom encountered in industrial processes; most are irregular in shape and the process always has to operate with a range of particle sizes simultaneously. This raises problems.

(i) How to characterize the shape of particles.
(ii) How to decide a suitable value, relevant to fluidized and packed beds, for the mean size of a particle within a batch of different particle sizes.

A non-spherical particle may have its shape quantified by defining a 'sphericity', φ:

$$\varphi = \frac{\text{surface area of a sphere of the same volume as the particle}}{\text{surface area of the particle}}. \quad (2.1)$$

(Notice that the sphericity φ is a non-dimensional quantity and is not the same quantity as the surface area per unit volume, S_v. The volume of a non-spherical particle, v_p, is given by

$$v_p = \frac{\pi d_p^3}{6} \quad (2.2)$$

where d_p is defined as

d_p = diameter of a sphere having the same volume as the particle. (2.3)

Thus, the surface area of the particle, a_p, is given by

$$a_p = \frac{\pi d_p^3}{6} S_v. \quad (2.4)$$

Hence, the sphericity φ is

$$\varphi = \frac{6}{d_p S_v}. \quad (2.5)$$

16 Particles and Fluidization

British Standard BS 4359 (1970) Part 3 provides data on measured values of the surface : volume ratio S_v for several types of particle. Many common granular types of particle have sphericities within the range 0.6–0.95. Surface area measurement is a complex matter requiring equipment not normally available outside of a laboratory, so that data on sphericity have to be obtained from tables in handbooks, or other sources. Table 2.1 below, however, gives some typical values.

Table 2.1 Typical sphericities of particles.

Particle	Sphericity
Sand (round)	0.92–0.98
Coal (crushed)	0.8–0.9
Mica flakes	0.28
Alumina	0.3–0.8
Catalysts	0.4–0.9
Limestone	0.5–0.9

When the value of sphericity required is critical, however, it is necessary to determine the sphericity of the actual particles being used, rather than rely upon general tables. Allen (1981) illustrates the considerable complexity of shape determination and alternative criteria. There is a wide variation in the sphericity of given materials because of the variety of shapes occurring naturally or after crushing. Generally, process plant has to operate with a range of particle sphericity and sizes and allowance for the variation of both from one batch of particles to the next has to be made by the plant designer.

Different criteria may be used to define the shape and size of a single irregular shaped particle, but equations (2.3)–(2.5) are sufficient here. With fluidized or packed beds, though, one is always having to deal with a *mixture* of particles of different sizes and, unless the particles are very large (>5 μm) or very small (<40 μm), the usual method used to determine particle size is the obvious and simple one of taking a small quantity of the particles from the bed and then, using a standard set of sieves (screens) of different aperture sizes, to make a size analysis—see figure 2.1. A discussion of other methods is beyond the scope of this book and the reader is referred to Allen (1981) for details. If the fraction of the mass of the sample retained by a particular aperture size, x_i, is then measured, the data can be processed in accordance with equation (2.6) below to obtain the mean particle size:

$$d_m = \left(\sum(x_i/d_i)\right)^{-1} \qquad (2.6)$$

where d_i is the arithmetic mean of two adjacent sieve (screen) aperture sizes.

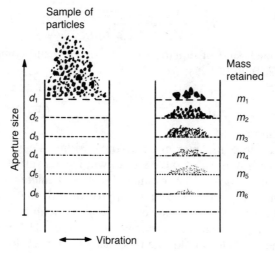

Figure 2.1 Sieve analysis using a standard set of sieves.

Equation (2.6) arises because for flow of fluid through a bed of particles, the pressure drop across the bed is of prime importance; hence the surface area of the particles is very relevant. Thus, with a mixture of particle sizes in the bed, the mean particle size d_m should be defined such that for a given volume of solids, the surface area:volume ratio calculated from the mean particle size is the same as that of the mixture. If, for simplicity, the mixture is made up of spherical particles and n_i is the number of particles of size d_i in the mixture, then the volume of solids

$$V_s = \sum \left(n_i \frac{\pi}{6} d_i^3 \right) \tag{2.7}$$

the surface area of particles

$$A_s = \sum (n_i \pi d_i^2) \tag{2.8}$$

and the mass fraction of particles of size d_i

$$x_i = \frac{n_i (\pi/6) d_i^3}{V_s}. \tag{2.9}$$

Now

$$\frac{x_i}{d_i} = \frac{\pi n_i d_i^2}{6 V_s}. \tag{2.10}$$

If, then, N_m particles of size d_m are required to make up a volume V_s and a surface area A_s of solids then the surface:volume ratio S_v is given by

$$S_v = \frac{A_s}{V_s} = \frac{N_m \pi d_m^2}{N_m (\pi/6) d_m^3} = \frac{6}{d_m}. \qquad (2.11)$$

If we sum each side of equation (2.10) for all i and compare this with equations (2.8) and (2.11):

$$\sum \left(\frac{x_i}{d_i}\right) = \left(\pi \sum (n_i d_i^2)\right)(6V_s)^{-1} = \frac{A_s}{6V_s}. \qquad (2.12)$$

Hence equation (2.6).

Spheres enclose the maximum volume within a given surface area; thus a spherical particle of sieve size d_i will have a surface area πd_i^2. For non-spherical particles, the estimated mean diameter d_i, determined from sieving, will have a volume approximately that of a sphere of diameter d_i, namely $\pi d_i^3/6$. However, from the definition of sphericity φ (equation (2.1)) the surface area will be $\pi d_i^2/\varphi$. The volume : surface ratio of the particle would then be $\varphi d_i/6$. Hence, for a mixture of non-spherical particles, all having the same shape factor φ, the mean particle size d_m using equation (2.6) would then be φd_m.

Equation (2.6) should however *not* be used if the sample of particles has a size distribution which is discontinuous (see figure 2.2(a) in which there are a number of peaks). A mean size of particle for such mixtures has little meaning. Beds of particles with such a distribution would, in any case, tend to segregate or mix non-uniformly on fluidizing.

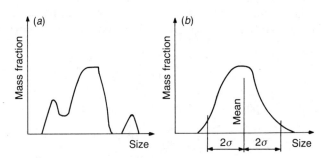

Figure 2.2 Types of size distribution. (a) Multi-modal. (b) Uni-modal.

An example of the use of equation (2.6) will illustrate how to proceed to obtain the various d_i from adjacent sieve (screen) aperture sizes and arrive at a mean particle size. Note that sieves (screens) have finite intervals between aperture sizes and that in the case of irregular shaped particles, the particle will pass through the aperture when it is presented to the aperture with its second largest diameter in the plane of the aperture; e.g. consider a long, thin particle compared with a spherical one. This second largest diameter is not the same as d_p defined by equation (2.3), but sieving is a simple, practical way of determining particle size and is sufficient for most practical purposes. However, Allen (1981) gives details of more refined methods.

Example 2.1. The following sieve analysis was obtained from a sample of particles. Determine the mean particle size.

Table 2.2 Sieve analysis of a sample of particles.

Sieve aperture size (μm)	Mass of particles retained in sieve (g)
1000	None
850	5.2
710	7.9
600	20.4
500	48.7
355	35.1
300	18.6
250	14.5
212	6.1
180	4.9
125	None
Total mass of sample (g)	161.4

First draw up a table of the mean adjacent sieve aperture size d_i and mass fraction x_i, as below.
Thus

$$d_1 = \tfrac{1}{2}(1000 + 850) = 925 \ \mu m$$
$$d_2 = \tfrac{1}{2}(850 + 710) = 780 \ \mu m$$
$$x_1 = 5.2/161.4 = 0.0322$$
$$x_2 = 7.9/161.4 = 0.0489$$

It is also useful to plot the data as shown in figure 2.2 to see if there are serious discontinuities or other peculiarities. Figure 2.3 shows a plot of x_i versus d_i from table 2.3. Using equation (2.6) gives the mean particle size

$$d_m = 1/234.372 \times 10^{-5} = 427 \ \mu m.$$

Table 2.3 Mass fractions of particles of given sizes and the determination of $\Sigma(x_i/d_i)$.

Aperture size range (μm)	d_i (μm)	x_i	x_i/d_i
850–1000	925	0.0322	3.481×10^{-5}
710–850	780	0.0489	6.269×10^{-5}
600–710	655	0.1264	19.298×10^{-5}
500–600	550	0.3017	54.855×10^{-5}
355–500	428	0.2175	50.818×10^{-5}
300–355	328	0.1152	35.122×10^{-5}
250–300	275	0.0898	32.655×10^{-5}
212–250	231	0.0378	16.364×10^{-5}
180–212	196	0.0304	15.510×10^{-5}

$$\sum \frac{x_i}{d_i} = 234.372 \times 10^{-5} \, \mu\text{m}^{-1}$$

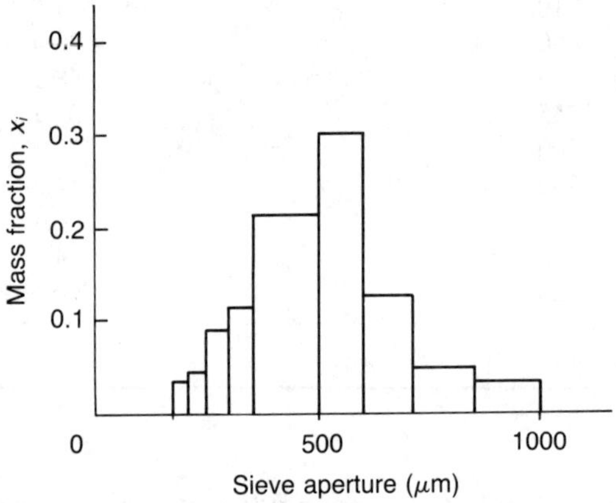

Figure 2.3 A plot of data from table 2.3.

2.1.2 Particle size range

Table 2.2 illustrates that all the particles in the sample are larger than those which could pass through a sieve aperture 180 μm square, but smaller than those which would *not* pass through the 1000 μm aperture.

Thus, the size range of the sample, r, is

$$r = 1000 - 180 = 820 \, \mu\text{m}. \quad (2.13)$$

This simple way of stating size range is sometimes adequate but any statistician would point out that it would be better to have a parameter which is less likely to be upset by extreme values (Moroney 1960).

Physical Properties of Solid Particles 21

Several different measures of dispersion, such as standard deviation and interquartile range, can be found in books on statistics. If the distribution is fairly symmetrical and uni-modal, such as that shown in figure 2.2(b), then 95% of the distribution of particle size would be within two standard deviations (2σ) from the mean size as indicated in figure 2.2(b) and 99% within three standard deviations (3σ). However, there does not seem to be any entirely satisfactory way of defining how wide a size distribution is, nor for comparison of such with that of a sample of particles having a different mean size. One common practice is to use a plot of cumulative percentage undersize versus sieve aperture size and define a relative size range R as the non-dimensional quantity

$$R = \frac{\frac{1}{2}(d_{0.84} - d_{0.16})}{d_m} \tag{2.14}$$

where $d_{0.84}$ is the particle size below which 0.84 of the mass of the sample lies, $d_{0.16}$ is the particle size below which 0.16 of the mass of the sample lies and d_m is the particle mean size $[\Sigma(x_i/d_i)]^{-1}$.

Table 2.4 gives the data for the cumulative plot shown in figure 2.4. Values taken from the plot are $d_{0.84} = 630$ μm and $d_{0.16} = 310$ μm. The mean size of the particle, d_m, already found in Example 2.1, is 427 μm. The relative size range is therefore

$$R = \frac{\frac{1}{2}(630 - 310)}{427} = 0.375.$$

Notice, also, that 68% of the sample is included in a size range of ± 160 μm about the mean size, whereas the size range which includes all extremes, equation (2.13), is ± 410 μm. Thus, it is sensible to appreciate the limitations of various measures of size range because this may affect the use to which such data can be put safely.

Table 2.4 Mass fractions of particles passing through sieve apertures.

Sieve aperture (μm)	Mass fraction passing through sieve aperture
1000	1.0
850	0.968
710	0.919
600	0.792
500	0.491
355	0.273
300	0.158
250	0.0682
212	0.0304
180	0

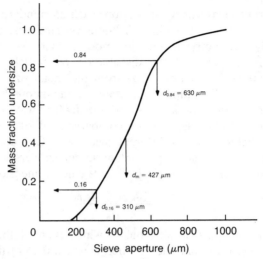

Figure 2.4 The cumulative mass fraction passing through a sieve aperture (table 2.4).

Size range is important since it is related to the voidage of a packed bed. If the size range is large, then the finer particles can fit into the spaces between the large particles, decreasing the voidage of the bed as a whole. Further, the finer particles can of course be transported from the bed by the fluidizing gas, but their presence in the voids can also affect the behaviour of the bed significantly; these features will be discussed later in §2.6.4.

2.1.3 Surface area of particles in a bed

One of the great advantages of particulate beds for carrying out gas-to-solid reactions is the potentially very large surface area of solid which is exposed to the gas. For example, Botterill (1975) has drawn attention to the fact that a cubic metre containing spheres each of diameter 100 μm has a total surface area of particles of the order of 30 000 m^2, a surface area of similar order to that of the Great Pyramid of Cheops in Egypt (Botterill 1975)!

From equation (2.1), the surface area of a single, non-spherical particle, a_i, of sieve size d_i will be

$$a_i = \frac{\pi d_i^2}{\varphi} \tag{2.15}$$

and for a bed of particles containing a mixture of sizes of similar shaped particles the total surface area of particles, A_s, will be

$$A_s = \frac{1}{\varphi}\sum(n_i \pi d_i^2) \tag{2.16}$$

which, by comparison with equation (2.11), leads to

$$A_s = \frac{6V_s}{\varphi d_m}. \qquad (2.17)$$

Example 2.2. The solid density of a material is 2640 kg m^{-3}. A bed of particles of this material has a mass of 2000 kg. The mean particle size is 725 μm and the particle shape is such that the particle sphericity is 0.85. Determine the total surface area of particles in the bed.

The volume of solids

$$V_s = \frac{\text{mass}}{\text{density}}$$

$$= \frac{2000}{2640} \text{kg} \, \frac{\text{m}^3}{\text{kg}}$$

$$= 0.758 \text{ m}^3$$

and, thus, the surface area of solids

$$A_s = \frac{6 \times 0.758 \text{ m}^2}{0.85 \times 725} \left(\frac{10^6 \, \mu\text{m}}{\text{m}} \right)$$

$$= 7380 \text{ m}^2.$$

2.1.4 Bed voidage

The voidage of a bed of particles, ε, is the fraction of the bed volume which is occupied by the space between the solid particles. Its value depends upon the shape of the particles, the pattern in which they are arranged in the bed, the size range of the particles (fine particles can fill the voids between the larger ones), the size of the bed (the voidage near the containing walls or immersed surfaces is different from that in the middle of the bed), etc. Voidages predicted from the geometry of single particles are therefore unreliable in practice. It is more reliable to determine voidage experimentally, for example, by comparison of the bulk density of a bed of particles, ρ_b, with the density of the solid of which the particles are made, ρ_p.

Example 2.3. Particles, loosely packed, are contained to a depth of 0.6 m in a vessel of 1.2 m diameter. The mass of the particles is 1100 kg and the density of the solid material is 2780 kg m^{-3}. Determine the voidage of the bed.

The volume of the bed

$$= \pi/4 \times 1.2^2 \text{ m}^2 \times 0.6 \text{ m} = 0.679 \text{ m}^3$$

the bulk density

$$\rho_b = \frac{1100 \text{ kg}}{0.679 \text{ m}^3} = 1621 \text{ kg m}^{-3}$$

and, thus, the voidage ε† is

$$\varepsilon = \left(1 - \frac{\rho_b}{\rho_p}\right) \quad (2.18)$$

$$= \left(1 - \frac{1621}{2780}\right)$$

$$= 0.417.$$

It is worth commenting that if the bed was tightly packed, or rammed down, the voidage would be smaller than that calculated above. Note, also, that a determination of particle size is not needed in the above case.

2.2 Classification of Particles According to Fluidization Characteristics

2.2.1 General remarks

Not all types of particle can be fluidized satisfactorily; some can be, but beds of different kinds of particles can behave differently when fluidized. Thus, some system for categorizing particles according to the way they behave during fluidization is desirable. For brevity, here, Geldart (1972) argued that the mean size on a surface area:volume ratio basis and the particle density are the important factors. A comparatively small mass of fine particles (say < 45 μm) in a batch of multi-sized particles can influence bed behaviour and, through the mathematics of mean particle size determination (equation (2.6)), can exert a considerable influence on the magnitude of the mean particle size d_m. Indeed, later work (Abrahamsen and Geldart 1980) showed that the actual proportion of particles smaller than about 45 μm was of great influence. The significance of this can be appreciated in relation to the classification

---

† $\quad$ Voidage, $\varepsilon = \dfrac{\text{volume of bed} - \text{volume of solids}}{\text{volume of bed}}$

$$= 1 - \frac{\text{volume of solids}}{\text{volume of bed}}$$

$$= 1 - \frac{m_p}{\rho_p} \times \frac{\rho_b}{m_b}$$

where m and ρ are the mass and density, respectively. However, as the voids may be treated as empty spaces, i.e. $m_p = m_b$, then

$$\varepsilon = 1 - \frac{\rho_b}{\rho_p}.$$

suggested by Geldart in 1973, which is based upon visual observations of bed behaviour at ambient conditions. The resulting classification relates to the influence of particle mean size and particle density on bed behaviour, and is shown in figure 2.5.

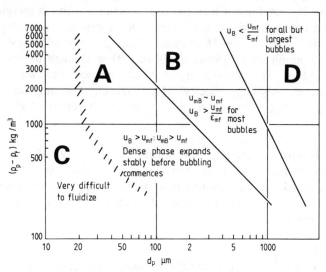

Figure 2.5 A powder classification diagram. (From Geldart 1972.) Reproduced by permission of Elsevier Sequoia S A.

2.2.2 Categories of particles

These fall into four categories: A, B, C and D. For ease of discussion they will be discussed in the order C, A, B, D.

Category C

Inspection of figure 2.5 shows that category C particles—generally of mean size $<30~\mu\mathrm{m}$ and/or of low density so that interparticle forces have a larger influence than the gravitational force on the particles—are extremely difficult to fluidize. With such particles the fluidizing gas tends to open up low-resistance channels through the bed and, once this has happened, most of the gas flows through such channels so that the distribution of the gas is far from uniform. Hence, the bed will never become properly fluidized. Mechanical stirring in the region close to the distributor can alleviate this unsatisfactory situation and improve the uniformity of gas distribution, but the extent of the improvement is limited.

Category A

Referring to figure 2.5, category A particles are generally within the size range 20–100 μm and have a particle density of less than about 1400 kg m^{-3}. They are often referred to as powders and beds of them expand considerably with increase in gas fluidizing velocity when the minimum fluidizing velocity U_{mf} is first exceeded. Expansion can then proceed until the fluidizing velocity reaches perhaps as much as two or three times U_{mf}. A further increase in fluidizing velocity is accompanied by the bed collapsing back to a magnitude of expansion corresponding to about that at minimum fluidization. The gas flow in excess of that required for minimum fluidization will then flow through the bed in the form of bubbles. The gas velocity at which the bed collapses is termed the 'minimum bubbling velocity', U_{mb}. During the quiescent period before the bed begins to bubble, particle mixing is very limited.

Category B
Category B particles are generally within the size range 40–500 μm and the density range 1400–4500 kg m^{-3}. Beds of these particles exhibit much less stable bed expansion than with category A particles. Free bubbling occurs when the minimum fluidizing velocity (or just slightly greater) is reached.

Category D
Category D particles are generally of mean size greater than 600 μm and/or of higher density than other categories. They require higher gas velocities to fluidize the bed than other categories, so much so that the gas flow regime through the voids surrounding the particles, required to fluidize the bed, becomes transitional or turbulent rather than laminar, while the way in which bubbles coalesce to form fewer but larger bubbles is different from those of categories A and B. The net result is that category D particles do not mix so well in the freely bubbling condition as those of categories A and B. This will be discussed further in §2.5.1.

Other system of classification
Recently, Grace (1986) has proposed a phase diagram to identify, broadly, flow regimes appropriate to combinations of gas velocity and particle properties. This is also discussed briefly in Geldart (1986). However, Geldart's system above is widely used at present and discussion will be terminated here.

It is timely now to return to the behaviour of beds of particles.

2.3 Pressure Drop Across Packed Beds

2.3.1 Flow through packed beds
It has already been remarked in §1.3.2 that the pressure drop across a

fluidized bed is given by equation (1.4) or its approximation, equation (1.5), and that the pressure drop is constant as the gas velocity increases from that at minimum fluidization to that at which significant particle entrainment occurs, as shown in figure 1.7. Ergun (1952) and Ergun and Orning (1949) refined a general equation which related pressure drop to gas flow for fixed beds, substituting the bed weight per unit base area for pressure drop at the minimum fluidizing velocity U_{mf}. It was assumed that the bed of particles was equivalent to a group of straight, parallel channels of equal size, such that the total surface area of the channels and the volume of fluid enclosed by them were equal to the total surface area of the particles and the volume of the voids, respectively, of a randomly packed bed.

It was argued initially that the pressure drop was the sum of two effects, namely that required to overcome viscous shear stresses (proportional to velocity) and that required to overcome kinetic energy losses (proportional to the square of the velocity).

However, the flow pattern through a fixed bed of particles is more complicated than that through a series of straight, parallel channels because the fluid flow paths through the voids are never straight, but have frequent changes in direction and repeated convergence and divergence of stream lines. Such complex geometrical boundaries create considerable difficulties for mathematical modelling of the flow. A detailed discussion is quite beyond the scope of this book; the reader is referred to Scheidegger (1974).

2.3.2 Modelling of the flow

The approach used by Ergun and Orning (1949) was nevertheless to model a packed bed as a series of identical, straight, parallel channels, and then to form an equation of the form, pressure gradient

$$\frac{dp}{dx} = av + b\rho v^2 \tag{2.19}$$

where v is the fluid velocity through the channels and a and b are coefficients. The first and second terms in equation (2.19) are subsequently multiplied by the dimensionless correlation factors α and β, respectively. Thus

$$\frac{dp}{dx} = \alpha av + \beta b\rho v^2. \tag{2.20}$$

Values of α and β were determined by conducting experiments.

The coefficient a was obtained from the well known Hagen–Poiseuille equation for pressure drop, Δp_v, over length, L, of a single straight tube of circular cross section of diameter d, in which the flow is entirely laminar; thus

28 Particles and Fluidization

$$\frac{\Delta p_v}{L} = \frac{32\mu_f v}{d^2} \qquad (2.21)$$

where v is the mean fluid velocity through the tube and μ_f is the viscosity.

The pressure drop Δp_k due to dissipation of kinetic energy in eddies or turbulence will be

$$\frac{\Delta p_k}{L} = \tfrac{1}{2}\rho_f v^2 \frac{f}{L} \qquad (2.22)$$

where f is a dimensionless friction factor, taken in this case as being equal to L/d.

The total pressure drop Δp_b over length L is therefore

$$\frac{\Delta p_b}{L} = 32\mu_f \frac{v}{d^2} + \frac{1}{2d}\rho_f v^2. \qquad (2.23)$$

If the bed is considered to be composed of N such tubes in parallel, then their length L and diameter d can be expressed in terms of the surface area and volume of solid particles in the bed and the bed voidage.

The surface area of the tube walls

$$A_w = N\pi d L \qquad (2.24)$$

and the volume of fluid in the tubes

$$V_f = \frac{N\pi d^2 L}{4}. \qquad (2.25)$$

Thus

$$\frac{\text{surface area of tube walls}}{\text{volume of fluid in the tubes}}, \frac{A_w}{V_f} = \frac{4}{d}. \qquad (2.26)$$

If the bed of particles is of depth L, containment diameter D and voidage ε then the surface area of particles

$$= \Sigma(n_p s_p) \qquad (2.27)$$

where n_p and s_p are the number and surface area of particles of each size, p, in the bed, the volume of solid particles

$$= \Sigma(n_p v_p) = (1-\varepsilon)\frac{\pi}{4}D^2 L \qquad (2.28)$$

where v_p is the volume of each particle of each size and the volume of voids

$$= \varepsilon\frac{\pi}{4}D^2 L. \qquad (2.29)$$

Pressure Drop Across Packed Beds

Now the surface-to-*void* volume of the bed, S_{vv}, is to be the same as that of the cluster of tubes, i.e.

$$\left(4\sum(n_p s_p)\right)(\varepsilon \pi D^2 L)^{-1} = \frac{4}{d}. \tag{2.30}$$

However, from equation (2.28)

$$\pi D^2 L = \left(4\sum(n_p v_p)\right)(1 - \varepsilon)^{-1}. \tag{2.31}$$

Inserting this into (2.30) gives

$$d = \left(\frac{4\varepsilon}{1 - \varepsilon}\right) / (\text{surface:volume ratio } S_v \text{ of the } particles) \tag{2.32}$$

$$= \left(\frac{4\varepsilon}{1 - \varepsilon}\right)\frac{1}{S_v}. \tag{2.33}$$

Now the fluid velocity v, through the voids, is related to a superficial fluidizing velocity U by

$$v = \frac{U}{\varepsilon}. \tag{2.34}$$

Substituting for v and d in equation (2.23) gives

$$\frac{\Delta p_b}{L} = \frac{2(1 - \varepsilon)^2}{\varepsilon^3}\mu_f S_v^2 U + \frac{1}{8}\frac{(1 - \varepsilon)}{\varepsilon^3}S_v \rho_f U^2. \tag{2.35}$$

Insertion of the dimensionless correlation factors α and β gives

$$\frac{\Delta p_b}{L} = 2\alpha \frac{(1 - \varepsilon)^2}{\varepsilon^3}\mu_f S_v^2 U + \frac{\beta}{8}\frac{(1 - \varepsilon)}{\varepsilon^3}S_v \rho_f U^2. \tag{2.36}$$

Ergun (1952) took the matter further, pointing out that it is customary to use a particle mean size d_m in pressure drop calculations. For spherical particles

$$d_m = \frac{6}{S_v}. \tag{2.37}$$

Substitution of this into equation (2.36) gives

$$\frac{\Delta p_b}{L} = 72\alpha \frac{(1 - \varepsilon)^2}{\varepsilon^3}\frac{\mu_f U}{d_m^2} + \frac{3\beta}{4}\frac{(1 - \varepsilon)}{\varepsilon^3}\rho_f \frac{U^2}{d_m}. \tag{2.38}$$

(If the particles are non-spherical and are of sphericity φ and mean size d then d_m in equation (2.38) is replaced by the product φd.)

Dividing each side of the equation (2.38) by $(1 - \varepsilon)^2 \mu_f U / \varepsilon^3 d_m^2$ gives

$$\frac{\Delta p_b d_m^2 \varepsilon^3}{L\mu_f U(1 - \varepsilon)^2} = 72\alpha + \frac{3\beta}{4}\frac{1}{(1 - \varepsilon)}\frac{\rho_f U d_m}{\mu_f} \tag{2.39}$$

where $\rho_f U d_m / \mu_f$ is the particle Reynolds Number Re_p. Ergun plotted a large amount of data from experiments with different types and sizes of particle and different fluids using equation (2.39). These plots led to values of 150 for 72α and 1.75 for $3\beta/4$, so that equation (2.39), when rearranged for non-spherical particles of sphericity φ, becomes

$$\frac{\Delta p_b}{L} = 150 \frac{(1-\varepsilon)^2}{\varepsilon^3} \frac{\mu_f U}{(\varphi d_m)^2} + 1.75 \frac{(1-\varepsilon)}{\varepsilon^3} \frac{\rho_f U^2}{\varphi d_m}. \qquad (2.40)$$

Equation (2.40) is commonly referred to as the Ergun equation.

It will be noted that the first term of the Ergun equation is linear in U and this will be dominant when the flow in the voids is laminar. The second term relates to turbulence. The Reynolds number of the flow decides whether the flow is laminar, turbulent or in transition. Botterill et al (1982) and also Ergun's plots (Ergun 1952) indicated how the Reynolds number Re, based upon particle diameter,

$$Re_p = \left(\frac{\rho_f U d_m}{\mu_f}\right) \qquad (2.41)$$

defines the flow regime in the voids (see table 2.5). (Note, also, that the Ergun equation neglects the effect of the static head of fluid, which may be significant for flow of liquid or high-pressure gas flowing vertically upwards through the bed.)

Table 2.5 The relationship between flow regime in voids and the particle Reynolds number.

$Re_p < 10$	Laminar flow in voids
$Re_p > 10^3$	Turbulent flow in voids
$10 < Re_p < 10^3$	Transition-to-turbulent flow in voids

2.4 Minimum Fluidizing Velocity and its Determination

2.4.1 Theoretical approach

If the pressure drop across a packed bed, given by the Ergun equation (2.40), is equated to that when fluidized, and solved for velocity U, then the solution would be the minimum fluidizing velocity U_{mf}. This approach was suggested by Ergun and Orning (1949); but careful note must be taken of the fact that when the pressure drop becomes equal to the weight of the bed per unit cross sectional area, the bed begins to expand, so that the voidage is different from the initial packed bed value. Further, the functions $(1-\varepsilon)/\varepsilon^3$ and $(1-\varepsilon)^2/\varepsilon^3$ in equation (2.40) are sensitive to very small changes in ε; hence, the subsequent solution for U depends upon having an accurate value for the bed voidage ε_{mf} at the minimum fluidizing velocity. Inserting

$(1 - \varepsilon_{mf})(\rho_p - \rho_f)g_n L$ (where g_n is the normal gravitational acceleration) for Δp_b, ε_{mf} for ε, and U_{mf} for U in equation (2.40) gives

$$(1 - \varepsilon_{mf})(\rho_p - \rho_f)g_n = 150 \frac{(1 - \varepsilon_{mf})^2}{\varepsilon_{mf}^3} \frac{\mu_f U_{mf}}{(\varphi d_m)^2} + 1.75 \frac{(1 - \varepsilon_{mf})}{\varepsilon_{mf}^3} \frac{\rho_f U_{mf}^2}{(\varphi d_m)}. \quad (2.42)$$

Multiplying each side by $\rho_f d_m^3 / \mu_f^2 (1 - \varepsilon_{mf})$ gives

$$\frac{\rho_f(\rho_p - \rho_f)g_n d_m^3}{\mu_f^2} = 150 \frac{(1 - \varepsilon_{mf})}{\varphi^2 \varepsilon_{mf}^3} \frac{\rho_f U_{mf} d_m}{\mu_f} + \frac{1.75}{\varphi \varepsilon_{mf}^3} \frac{\rho_f^2 U_{mf}^2 d_m^2}{\mu_f^2}. \quad (2.43)$$

Now the left-hand side of equation (2.43) is the dimensionless group known as the Archimedes Number, Ar:

$$Ar = \frac{\rho_f(\rho_p - \rho_f)g_n d_m^3}{\mu_f^2} \quad (2.44)$$

which is

$\tfrac{3}{4} \times$ drag coefficient $\times Re_{mf}^2$.

It will be noted that the particle Reynolds number (equation (2.41)), based on the minimum fluidizing velocity U_{mf}, appears on the right-hand side of equation (2.43). Thus, at minimum fluidizing velocity

$$Ar = 150 \frac{(1 - \varepsilon_{mf})}{\varphi^2 \varepsilon_{mf}^3} Re_{mf} + \frac{1.75}{\varphi \varepsilon_{mf}^3} Re_{mf}^2. \quad (2.45)$$

Theoretically, if sufficiently accurate values for ε_{mf} and the mean diameter of a particle d_m could be obtained, then equation (2.45) could be used to calculate the minimum fluidizing velocity U_{mf}, expressed in terms of Ar and Re_{mf}. Botterill et al (1982) have conducted experiments, however, which showed that ε_{mf} varies with bed temperature in a complex manner and is not easily predicted when particles are in the size and density ranges of 40–500 μm and 1400–4000 kg m^{-3}, respectively. On the other hand, larger, denser particles, for which $Ar \geqslant 26\,000$ and $Re_{mf} \geqslant 12.5$, do not seem to show an increase in ε_{mf} with bed temperature. Problems of predicting the minimum fluidizing velocity thus remain.

In practice, by far *the best procedure is to make a direct experimental measurement* of pressure drop across the bed at gradually decreasing gas velocity, plotting the data in a similar manner to that shown in figure 1.7, and reading off the minimum fluidizing velocity. Ideally, because of the variation of voidage with temperature referred to earlier, the bed temperature should be that at which the plant would have to operate.

If, however, one is forced to make an estimate of minimum fluidizing velocity without carrying out a pressure drop/fluidizing velocity experiment, probably the best that can be done is to obtain, as accurately as possible, an estimate of bed voidage at minimum fluidizing velocity, and then employ the Ergun equation (2.43) or (2.45). Such an estimate of voidage might be made by determining the bulk density ρ_b of a *loosely packed* bed by pouring a quantity of particles gently into a measuring cylinder of known volume and measuring the mass of particles. The particle mean size d_m can be found from sieve analysis and an estimate made of sphericity φ from the known shape of the particles; the fluidizing gas density ρ_f and viscosity μ_f can be found from tables of properties.

An example may make the procedure clear.

Example 2.4. A bed of particles of mean size 427 μm was found to have a density when loosely packed of 1620 kg m^{-3}; the density of the individual particles was 2780 kg m^{-3} and their sphericity was 0.73. Calculate the minimum fluidizing velocity when fluidized by air under ambient conditions, where the air density is 1.21 kg m^{-3} and the viscosity is 1.82×10^{-5} kg m^{-1} s^{-1}.

The voidage

$$\varepsilon_{mf} = \left(1 - \frac{1620}{2780}\right) = 0.417.$$

Archimedes number

$$Ar = \frac{9.81 \times (427 \times 10^{-6})^3 \times 1.21 \times (2780 - 1.21)}{(1.82 \times 10^{-5})^2} = 7753.$$

The coefficients of Re_{mf} and Re_{mf}^2 in equation (2.45) are, respectively

$$\frac{150(1 - 0.417)}{0.73^2 \times 0.417^3} = 2263$$

$$\frac{1.75}{0.417^3 \times 0.73} = 33.06.$$

Equation (2.45) thus becomes the quadratic

$$33.06 \, Re_{mf}^2 + 2263 \, Re_{mf} - 7753 = 0$$

whose solutions are $Re_{mf} = -71.7$ or $+3.266$, of which only the positive root is physically admissible.

From equation (2.41) for Re_{mf}

$$U_{mf} = \frac{3.266 \times (1.82 \times 10^{-5})}{1.21 \times (427 \times 10^{-6})} = 0.114 \text{ m s}^{-1}.$$

2.4.2 Empirical correlations

Notice that if no measurement of voidage ε_{mf} or sphericity φ is available, the only method remaining is to use an empirical correlation.

The reader is warned, however, that values of U_{mf} calculated from empirical correlations are likely to give values very different from those obtained using the procedure above. As Botterill *et al* (1982) point out, this is because the authors have attempted to average out the values of φ and ε_{mf} over the range of particle sizes and shapes covered. For example, Wen and Yu (1966) noted that, for their particular range of materials

$$\frac{1-\varepsilon_{mf}}{\varphi^2 \varepsilon_{mf}^3} \simeq 11.0 \quad \text{and} \quad \frac{1}{\varphi \varepsilon_{mf}^3} \simeq 14.$$

Inserting these values into the Ergun equation (2.45) gives the quadratic equation

$$Ar = 1650\, Re_{mf} + 24.5\, Re_{mf}^2 \qquad (2.46)$$

which, when solved for Re_{mf}, yields

$$Re_{mf} = \frac{-1650 \pm (1650 + 4 \times 24.5\, Ar)^{1/2}}{2 \times 24.5} \qquad (2.47)$$

of which the positive root is

$$Re_{mf} = (33.7^2 + 0.0408\, Ar)^{1/2} - 33.7. \qquad (2.48)$$

Equation (2.48) is probably at its most applicable for particles larger than 100 μm.

Baeyens and Geldart (1973) proposed a different empirical correlation for particles >100 μm:

$$Ar = 1823\, Re^{1.07} + 21.7\, Re_{mf}^2. \qquad (2.49)$$

For smaller particles (<100 μm), a correlation due to Baeyens (1973), equation (2.50), appears to give the best agreement with experiment, according to Abrahamsen and Geldart (1980):

$$U_{mf} = \frac{9.4 \times 10^{-4}(\rho_p - \rho_f)^{0.934} g^{0.934} d_m^{1.8}}{\mu_f^{0.87} \rho_f^{0.066}}. \qquad (2.50)$$

Goroshko *et al* (1958) derived an expression from the Ergun equation, (2.45), but neglected the product term in the quadratic in so doing (see Botterill *et al* 1982). Thus, the Ergun equation (2.45) is a quadratic:

$$a\, Re_{mf}^2 + b\, Re - Ar = 0 \qquad (2.51)$$

where

$$a = \frac{1.75}{\varphi \varepsilon_{mf}^3} \qquad (2.52)$$

and

$$b = \frac{150(1 - \varepsilon_{mf})}{\varphi^2 \varepsilon_{mf}^3} \qquad (2.53)$$

34   Particles and Fluidization

in which φ is taken as 1.0.

The solution of equation (2.51) giving a positive root is

$$Re_{mf} = \frac{-b + (b^2 + 4aAr)^{1/2}}{2a}. \tag{2.54}$$

If the numerator and denominator of equation (2.51) are multiplied by $(b + \sqrt{b^2 + 4aAr})$ the solution can be expressed as

$$Re_{mf} = Ar\left\{\frac{b}{2} + \left[\left(\frac{b}{2}\right)^2 + aAr\right]^{1/2}\right\}^{-1}. \tag{2.55}$$

The quoted Goroshko equation is

$$Re_{mf} = \frac{Ar}{b + \sqrt{aAr}} \tag{2.56}$$

which is not the same as (2.55) because it wrongly implies that

$$\left(\frac{b}{2}\right)^2 + aAr = \frac{b}{2} + \sqrt{aAr}.$$

Thus, the denominator of Goroshko's equation (2.56) is larger than that of the solution to Ergun's equation given by equation (2.55) and so the value of Re_{mf} predicted from Goroshko is smaller than that predicted by Ergun in equation (2.54). The extent of the discrepancy depends upon the value of the Archimedes number Ar. Thus

$$\frac{Re_{mf}(\text{Ergun})}{Re_{mf}(\text{Goroshko})} = \frac{b + \sqrt{aAr}}{\{b/2 + [(b/2)^2 + aAr]^{1/2}\}}. \tag{2.57}$$

Taking, for example, in equations (2.55) and (2.56) a value for the voidage ε_{mf} of 0.4 and $\varphi = 1$, then $a = (1.75/0.4^3) = 27.3$ and $b = 150 \times 0.6/0.4^3 = 1406$, giving

$$\frac{Re_{mf}(\text{Ergun})}{Re_{mf}(\text{Goroshko})} = \frac{1406 + (27.3Ar)^{1/2}}{703 + (494\,209 + 27.3Ar)^{1/2}}. \tag{2.58}$$

Table 2.6 shows the right-hand-side changes with Archimedes number. It will be seen that the discrepancy between Ergun and Goroshko at the chosen voidage ε_{mf} is greatest where $10^4 < Ar < 10^5$.

It is fortuitous that the discrepancy tends to be compensated for if the change in bed voidage with temperature is neglected for beds of particles falling within Geldart's category B.

It is appropriate again to draw attention to the fact that most empirical correlations were formulated from experiments with beds at or near ambient temperature and to beware of placing much reliance upon the accuracy of extrapolations from such correlations to higher-temperature situations. An increase in bed temperature can result in a transition in bed behaviour from that typical of category D particles at ambient temperature to category B (see Botterill et al 1982), the boundary coming at $Ar \leq 26\,000$ and $Re_{mf} \leq 12.5$.

Table 2.6 The variation of the ratio of predicted Re_{mf} from equation (2.58) with Archimedes number Ar, taking $\varepsilon_{mf} = 0.4$ and $\varphi = 1$.

Archimedes number, Ar	$\dfrac{Re_{mf}(\text{Ergun})}{Re_{mf}(\text{Goroshko})}$
10	1.008
10^2	1.036
10^3	1.093
10^4	1.221
10^5	1.224
10^6	1.110
10^7	1.040

The shape factor can be estimated from a test at ambient temperature on flow through a packed bed (see §2.4.5).

2.4.3 Summary

Summarizing the above it will be seen that these theoretical methods, developed for predicting minimum fluidizing velocity, have serious limitations, particularly when voidage at minimum fluidization and particle sphericity are not known with great accuracy. It should be remarked that the voidage at minimum fluidization at ambient temperature is not necessarily the same as at elevated temperature. The reader is referred to Botterill *et al* (1982) for further details.

In general it is wisest to make a direct measurement of minimum fluidizing velocity, as described in §2.4.4.

2.4.4 Experimental determination of minimum fluidizing velocity

For this purpose a small fluidized bed, say 100–150 mm in diameter, is sufficiently large for most powders or granular materials. A static pressure probe used to measure the pressure drop across the bed is connected to a manometer immersed in the bed of particles, as shown in figure 2.6. A fan or air supply, an air flow meter and flow control valve are also required. A quantity of the particles should be placed in the container to a depth at which the bed might be expected to operate, or to about 100–150 mm. First, the bed of particles should be fluidized vigorously, to break down any packing of the particles, then the gas velocity should be decreased in increments, the pressure drop across the bed being recorded at each flow rate. The data should then be processed as in the following example.

Example 2.5. Table 2.7 gives bed pressure drop/fluidizing velocity data for a bed of granular material operating at ambient conditions. Estimate the minimum fluidizing velocity.

36   Particles and Fluidization

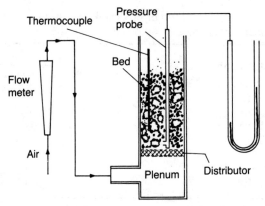

Figure 2.6 The experimental apparatus used for the determination of minimum fluidizing velocity.

Table 2.7 Bed pressure drop/gas velocity data at ambient temperature and pressure.

Gas velocity (m s^{-1})	Bed pressure drop (mm water gauge)
0.02	35
0.04	70
0.06	105
0.08	140
0.10	173
0.12	201
0.14	220
0.16	231
0.18	239
0.20	244
0.22	244
0.24	244
0.26	244
0.28	244
0.30	244

First, plot the data in table 2.7, as shown in figure 2.7, then extrapolate the straight line of the packed bed region of the graph and fluidized bed region until they intersect at point P, as shown. The fluidizing velocity at point P is then taken as the minimum fluidizing velocity U_{mf}, which is about 0.14 m s^{-1}.

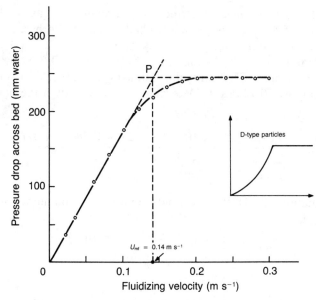

Figure 2.7 Pressure drop versus fluidizing velocity data from table 2.7.

Comment
It should be remarked that the shape of the pressure drop/fluidizing velocity graph in figure 2.7 differs from those shown in figure 1.7. The difference lies in the apparently gradual transition from the packed bed regime to a fluidized bed regime. Such a gradual transition usually occurs when the particle size range is wide. (If the particles were larger and denser, i.e. of D type, the shape of the graph in figure 2.7 during the packed bed region would be curved, as shown in the inset.)

Estimation of particle sphericity φ from bed pressure drop/fluidizing velocity data
The minimum fluidizing velocity U_{mf}, obtained as shown in Example 2.5, may also be used to estimate the sphericity φ of the particles if an estimate of the voidage ε_{mf} at minimum fluidization can be made. ε_{mf} may be determined by measuring the bulk density ρ_b of a loosely packed bed of the particles and the particle density ρ_p, and using equation (2.18), as in Example 2.3.

The Archimedes and Reynolds numbers, Ar and Re_{mf}, can then be calculated and substituted into equation (2.45) along with U_{mf}, the fluid properties μ_f and ρ_f and the particle mean size d_m. Equation (2.45) can then be solved for sphericity φ as follows.

Example 2.6. The particles referred to in Example 2.5 above have a mean size d_m of 427 μm. A loosely packed bed of them has a voidage ε_{mf} of 0.417. The fluidizing gas density and viscosity, ρ_f and μ_f, are 1.21 kg m^{-3} and 1.82 × 10^{-5} kg m^{-1} s^{-1}, respectively. The particle density ρ_p is 2780 kg m^{-3}. The minimum fluidizing velocity U_{mf} determined experimentally was 0.14 m s^{-1}. Determine the sphericity of the particles.

From equation (2.44)
$$Ar = \frac{9.81 \times (427 \times 10^{-6})^3 \times 1.21 \times (2780 - 1.21)}{(1.82 \times 10^{-5})^2}$$
$$= 7753.$$

From equation (2.41) the Reynolds number at minimum fluidizing velocity U_{mf} is
$$Re_{mf} = \frac{1.21 \times 0.14 \times (427 \times 10^{-6})}{1.82 \times 10^{-5}}$$
$$= 3.974.$$

Substituting values into the Ergun equation (2.45) leads to
$$7753 = \frac{150(1 - 0.417)}{0.417^3 \varphi^2} \times 3.974 + \frac{1.75}{0.417^3 \varphi} \times 3.974^2$$
$$7753\varphi^2 - 381.1\varphi - 4793 = 0$$

a quadratic equation whose roots are
$$= \frac{381.1 \pm [381.1^2 - 4 \times 7753 \times (-4793)]^{1/2}}{2 \times 7753}$$
$$= 0.811 \text{ or } -0.762.$$

Only the positive root, $\varphi = 0.811$, has physical meaning and hence the sphericity $\varphi = 0.811$.

2.5 Two-phase Theory of Fluidization, Bubbles and Fluidization Regimes

2.5.1 Two-phase theory

Section 1.3.2 described the broad sequence of events as the fluidizing gas velocity was increased progressively. The fluidized bed was regarded as a 'well behaved' one, i.e. the gas being distributed uniformly across the base of the bed, good mixing of particles, good contact between particles and fluidizing gas etc. The uniformity of the initial fluidizing gas distribution depends upon the distributor design, but mixing of particles and the degree of contact between the fluidizing gas and the surface of the particles are influenced strongly by the behaviour of the bubbles of gas rising through the bed. A bubbling action promotes good mixing of the particles, thereby helping to make the bed temperature uniform.

The two-phase theory (see, for example, Davidson and Harrison 1971, Kunii and Levenspiel 1969) assumes that all gas flow in excess of that required for incipient fluidization flows through the bed in the form of bubbles. This can be expressed by an equation such as (2.59) below.

The total volumetric gas flow

$$\dot{V}_g = U_{mf}A + (n_b v_b) \quad (2.59)$$

where A is the cross sectional area of the bed, n_b is the number of bubbles, each of volume v_b, per unit time passing through the bed at a given level and v_b is the volume of each bubble at that level.

The volumetric gas flow $\dot{V}_g$ can be written as the product of fluidizing velocity U and cross sectional area A and equation (2.59) can be rearranged to give the flow rate through the bubble phase, $n_b v_b$, as

$$n_b v_b = (U - U_{mf})A. \quad (2.60)$$

What equations (2.59) and (2.60) do *not* indicate, however, is the bubble size (volume) distribution, i.e. the number of bubbles per unit time of each size of bubble; neither do they indicate the velocity at which the bubbles rise through the bed, coalescence patterns etc. All these quantities exert considerable influence.

2.5.2 Bubble rise velocity

The velocity U_b at which a single, isolated bubble rises in a large-diameter vessel (a considerable simplification since in real situations bubbles often rise in swarms, coalesce or split, while in small vessels they are affected by the presence of the containing walls) is given by

$$U_b = 0.71\sqrt{gD_b} \text{ (m s}^{-1}) \quad (2.61)$$

where g is the gravitational acceleration and D_b is the bubble diameter, taken as the diameter of a sphere having the same volume as the bubble.

(An expression of this form has been deduced from a consideration of the irrotational motion of fluid past a sphere in an inviscid fluid (Rippin 1959). Actual experimental observations by Davies and Taylor (1950) yielded equation (2.61); see also Davidson and Harrison (1963).)

Should the bubble diameter D_b be larger than about 30% of the diameter of the bed containing vessel, then the single bubble rise velocity U_b has been found to be given by

$$U_b = 0.35\sqrt{gD_c} \quad (2.62)$$

where D_c is the diameter of the bed containing vessel.

This bubble rise velocity is the same as that to be found in the 'slug flow' regime described briefly in §2.5.4.

In most practical situations the fluidizing velocity is sufficiently greater than the minimum fluidizing velocity, U_{mf}, for bubbles to be formed continuously. The range of bubble sizes, shapes and rising velocities is considerable (see, for example, a recent paper by Glicksman *et al* 1985). Bubbles are often in sufficiently close proximity to each other that the flow field around each bubble is influenced by a neighbouring one; growth, coalescence and splitting of bubbles occur during their passage through the bed. These conditions are very different from the motion of a single isolated bubble through an incipiently fluidized bed. It follows that in the present state of knowledge, the distribution of bubble size is not predictable nor easily measured. Bubbles are of great importance because they influence particle mixing, gas-to-particle contacting, bed expansion, gas by-pass, elutriation etc.

It is highly desirable therefore to carry out experiments to establish the likely behaviour and performance of a bed rather than to try to predict its performance quantitatively. Unfortunately, such experiments may be costly and, even then, may have to be conducted on equipment of significantly smaller scale than the plant to be built commercially. Thus, the data from such experiments require critical appraisal before using them quantitatively to design a larger-scale plant.

However, in the absence of reliable experimental data, the best that can be done is to use a semi-empirical approach to estimate bubble velocity, accepting that the errors arising may be very large, and then to try to allow for the consequences of this in other parts of the design. An example of such an approach is equation (2.63) below, for bubble velocity in a bubbling bed

$$U_b = k(U - U_{mf}) + 0.71\sqrt{gD_b} \tag{2.63}$$

where k is a constant, which, in the region well above the distributor, for example, is approximately 1.0. Notice that equation (2.63) leads to much higher bubble velocities than the single isolated bubble velocity calculated from equation (2.61). If equation (2.63) is to be used to predict bubble velocity, the bubble size D_b must be known; the judgement of a suitable magnitude for D_b depends upon having observed bubbles with the particular type of bed being considered. In this connection, a recent, brief paper (Stergiou and Laguerie 1984) drew attention to the sensitivity of predictive models to bubble size and to events in the region near the distributor (which are influenced greatly by the type of distributor employed and the properties of the particles). The x-ray observation of bubbles in fluidized beds by Rowe *et al* (1984) and the probe measurement of size and velocities by Glicksman *et al* (1985) demonstrate the complexity of the variation of bubble behaviour, which serves to underline the difficulties of making predictions about bubble size and frequency.

Two types of bubbles can be identified with A- and B-type particles, namely those whose rise velocity is smaller than the gas velocity through the voids in the particulate phase of the bed, known as 'slow bubbles', and the converse, known as 'fast bubbles'. The slow bubbles give rise to a through flow of gas, as shown in figure 2.8(*a*), so that gas is exchanged between the bubble and particulate phases of the bed. If as shown in figure 2.8(*b*) the bubble rises faster than the gas velocity through the voids, then the gas within the rising bubble stays with it, makes regular excursions to and from the surrounding cloud, yet never leaves the cloud, as shown theoretically by Davidson (1961) and confirmed experimentally by Rowe (1964). The faster the bubble rises in relation to the gas velocity through the voids in the dense phase of the bed, the thinner the surrounding cloud becomes. The consequence of high bubble velocity is that the quantity of gas trapped within the bubble and the cloud passes through the bed with only the smallest amount of exposure of gas to the main body of particles; this is compounded furthermore because of the reduced bubble residence time. Hence for the greatest amount of gas-to-particle interaction, small, slowly rising bubbles are to be encouraged.

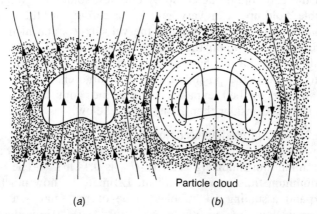

Figure 2.8 Gas flow patterns between bubble and particulate phases. (*a*) Slow bubble, $U_b < U_{mf}$. (*b*) Fast bubble, $U_b > U_{mf}$.

The terms 'slow' and 'fast' bubbles are, however, misleading when used in connection with the large, dense particles of Geldart's group D

42 Particles and Fluidization

type. With these, the gas velocity through the voids, required to fluidize the bed, is very much higher than the bubble rise velocity, so the bubbles would be classified as 'slow'. Yet, with such particles, the bubble rise velocity is greater than that of the 'fast' bubbles encountered with A- and B-type particles. Also, with beds of D-type particles, the gas flow through the voids is in the turbulent regime; thus, instead of bubble coalescence being caused by a bubble being caught up in the wake of a larger fast-rising bubble, as with A- and B-type materials, coalescence occurs cross-wise between bubbles at a similar level in the bed. Table 2.8 summarizes these points.

Table 2.8 Bubble categories in fluidized beds.

Particle category	Bubble types encountered
A	Both 'slow' ($U_b < U_{voids}$) and 'fast' ($U_b > U_{voids}$)
B	As for type A
C	Not able to be fluidized
D	'Slow', but value of U_b greater than with A and B particles. Bubble growth by cross-wise coalescence

2.5.3 Bed expansion

The presence of bubbles within the bed causes the bed to expand; the extent of the bed volume taken up by bubbles, sometimes referred to as the 'bubble hold-up', can vary greatly, according to the bubble size distribution within the bed, because the velocity U_b at which any given bubble rises depends upon the size of the bubble, the interaction with other bubbles etc. It is, however, useful, purely for gaining a little insight, to calculate the possible effect of a large change in bubble size on the amount of bed expansion using the admittedly over-simple equation (2.63). Consider then Example 2.7 below.

Example 2.7. A large-diameter fluidized bed reactor is to operate at a fluidizing velocity U of 0.31 m s^{-1}. Experiments with the particles show that their minimum fluidizing velocity U_{mf} is 0.14 m s^{-1}. The bed depth H_{mf} at minimum fluidization is 0.65 m. Estimate by how much the bed might expand assuming all bubbles to be of the same size, being of volume equivalent to a sphere of either (a) 0.03 m diameter or (b) 0.003 m diameter.

Part (a)
From equation (2.63), with $k = 1.0$, the bubble velocity

$$U_b = 1.0 \times (0.31 - 0.14) + 0.71 (9.81 \times 0.03)^{1/2}$$
$$= 0.555 \text{ m s}^{-1}.$$

Consider now figure 2.9(a), the 0.03 m diameter bubble situation. The time t_b for a bubble to rise through the bed

$$t_b = \frac{H_a}{U_b}. \tag{2.64}$$

The expansion of the bed due to the bubbles in it is

$$H_a - H_{mf} = \left(\frac{\text{gas flow through the bubble phase}}{\text{cross sectional area of the bed}}\right) t_b$$

which, from equations (2.63) and (2.64), yields

$$H_a - H_{mf} = (U - U_{mf})\frac{H_a}{U_b} \tag{2.65}$$

or

$$\frac{H_a - H_{mf}}{H_a} = \frac{U - U_{mf}}{U_b}. \tag{2.66}$$

Substituting values into equation (2.66) yields

$$\frac{H_a - H_{mf}}{H_a} = \frac{0.31 - 0.14}{0.555}$$

$$= 0.306$$

$$H_a = \frac{H_{mf}}{1 - 0.306}$$

$$= \frac{0.65}{0.694}$$

$$= 0.937 \text{ m}.$$

The bed expansion

$$(H_a - H_{mf}) = 0.937 - 0.65$$

$$= 0.287 \text{ m}.$$

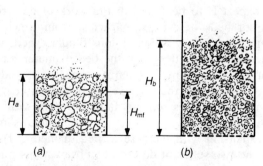

Figure 2.9 The effect of bubble size on bed expansion. (a) 0.03 m bubbles. (b) 0.003 m bubbles.

Part (b)
The bubble velocity

$$U_b = 1.0 \times (0.31 - 0.14) + 0.71(9.81 \times 0.003)^{1/2}$$
$$= 0.292 \text{ m s}^{-1}.$$

Using the nomenclature in figure 2.9 (b) yields

$$\frac{H_b - H_{mf}}{H_b} = \frac{0.31 - 0.14}{0.292}$$
$$= 0.582$$

giving

$$H_b = 1.556 \text{ m}.$$

The bed expansion

$$(H_b - H_{mf}) = 0.906 \text{ m}.$$

Thus, if the excess gas flow $(U - U_{mf})$ passed through the bed as small bubbles, the bed would be expected to expand more than if the bubble phase consisted of large bubbles, as depicted in figure 2.9. The smaller bubbles, case (b), would be expected to have a longer residence time in passing through the bed. Thus the residence time

$$t_b = \frac{0.937}{0.555} = 1.69 \qquad \text{for case } (a)$$

and

$$t_b = \frac{1.556}{0.292} = 5.33 \qquad \text{for case } (b).$$

The results of the above calculations should not be taken farther than above for the time being. Their limitation is in the fact that the bubble size and the velocity are changing as they rise through the bed.

A cautionary illustration of the uncertainty about the amount of bed expansion to expect is to be found in the work of de Groot (1967) on the expansion of beds of catalyst particles, from which figure 2.10 is derived. The percentages by which various fluidized beds were observed to expand have been plotted against the bed diameter for two different size ranges of particles, namely a narrow size range and a broad size range. It will be seen that the observed bed expansion when using the broad size range of particles remains fairly constant with bed diameter, while beds having particles of narrow size range show a marked reduction in expansion with increase in bed diameter. The only difference between the two series of de Groot's observations was the quantity of fine material in the bed and, thus, the mean particle size $(1/\Sigma(x_i d_i))$.

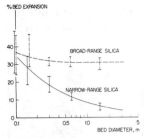

Figure 2.10 Bed expansion as a function of bed diameter, the only difference being the size distribution of catalyst; the expansion reflects bubble hold-up. (After de Groot 1967.)

The calculations made in Example 2.7 do not include any parameters which describe the bed diameter, particle size or size range and so they could not take any account of the very important phenomena demonstrated by de Groot.

This matter will be referred to again in §2.5.4.

2.5.4 Bubble growth and slugging

Bubbles are initiated near the distributor but there is, even now, no satisfactory explanation as to why they form. The pattern of gas flow in and out of bubbles (exchange of gas between particulate and bubble phases) can have an important effect on the degree of gas-to-solid contact. A discussion of bubble splitting and coalescence are beyond the scope of this section and the reader is referred to Rowe (1971); however, some comments on net effects may be helpful here.

Category A and B particles
With group B particles (§2.2.2), bubbles tend to grow in size by coalescence, as they pass through the bed, until their diameter is limited (provided the bed is sufficiently deep) by the diameter of the containment (see figure 2.11(*a*)). Group A particles, on the other hand, seem to exhibit a maximum stable bubble size within the upper reaches of deep beds, due to an apparent equilibrium between bubble growth by spontaneous coalescence and bubble splitting (figure 2.11(*b*)). The size of these bubbles seemed to be influenced by the fraction of fines within the particles (Botterill 1983).

The work of de Groot (1967) referred to in §2.5.3 and figure 2.10 illustrates this latter point, where the overall bed expansion brought about by the volume of bubbles within the bed depends upon whether or not the mixture of particles contains a significant fraction of fines. Such a fraction may be sufficient to depress the mean particle size of the mixture to such an extent that the bed tends to behave as one composed of category A particles, with a consequent large expansion of the bed.

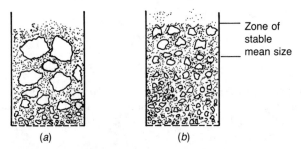

Figure 2.11 Bubble growth with group A (*b*) and B (*a*) particles.

Category D particles
Despite the growing importance of using beds of large particles in, for example, fluidized bed combustion systems, where there is the strongest economic incentive to operate at the highest practicable gas velocity to obtain high throughput or combustion intensity, relatively little experimental work with large particles has been reported. It has already been mentioned in §2.5.2 that the gas velocity through the voids in beds of category D particles exceeds that of bubbles and that bubble coalescence tends to be by horizontal interaction. There seems, according to Glicksman *et al* (1985), to be no observable maximum size to which bubbles in beds of D-type particles grow. Space does not permit the matter to be discussed further here, but the reader is referred to Yates (1983) and Cranfield and Geldart (1974).

Slugging
When bubbles grow to a size at which they occupy the whole cross sectional area of the containing vessel, the bed is said to be 'slugging'. Such a phenomenon can be demonstrated with a small-scale bed (say 40 mm in diameter) containing particles to a sufficient depth and operating at a sufficiently high gas velocity. These bubbles carry a slug of particles ahead of them, as shown in figure 2.12, until such time as the particles unlock from each other and fall back into the bed. Experiments with small-diameter beds are usually limited by the gas velocity at which slugging occurs, and this is normally a much smaller velocity than that at which a large-scale bed would have to operate (because of the high throughput required for commercial plant to satisfy economic criteria). Mixing under slugging conditions is very different from that occurring in a freely bubbling bed. Moreover, the overall

pressure drop across the bed will be greater than that required just to support the weight of bed particles per unit cross sectional area, as in equation (1.5), because energy has to be supplied to accelerate the slug of particles upwards; pressure fluctuations across the bed will also increase in amplitude. This does not necessarily mean to say that operating a fluidized bed in the slugging regime is something to be avoided at all costs, but that laboratory data obtained from small-scale equipment for scale-up of fluidized beds requires more cautious interpretation before it can be used to predict large-scale plant performance. However, the use of a small, slugging fluidized bed in model reactor studies has the merit of regular, stable bubble behaviour; the poor mixing and heat transfer, encountered when low gas velocity is employed as a means of limiting bubble formation, can thus be circumvented (Hovmand and Davidson 1971).

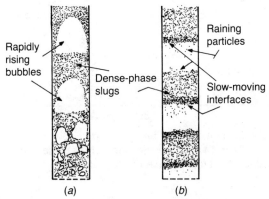

Figure 2.12 Alternative types of slugging behaviour after Hovmand and Davidson (1971).

2.6 Mixing, Elutriation and Transport of Solids

2.6.1 General mechanism of mixing of particles

When a single isolated bubble passes upward through a bed of incipiently fluidized particles, the particles tend to flow around it in a manner somewhat like that encountered when a body moves through a fluid stream. The bubble, like the body in the fluid stream, tends

(i) to be surrounded by a flow field pattern of a type which causes the particles immediately behind the bubble to follow it and

(ii) to develop a 'wake' behind it, in which particles are carried along with the bubble as if attached to it, except that some particles are exchanged with the bed along the path of the bubble. These particles in the wake become deposited at the free surface of the bed when the bubble breaks through the surface.

These features are illustrated in figure 2.13, which shows a 'streak' of particles transported up through the bed by the passage of the bubble. Particles which are brought from the lower part of the bed to the bed surface in this way must of course be replaced by other particles flowing from the rest of the bed into the vacated space; thus there is a general downward flow of particles in the region surrounding the path of the rising bubble in figure 2.13.

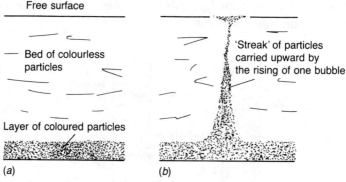

Figure 2.13 The upward transport of particles by a single bubble (after Rowe 1971). (*a*) The bed section before passage of a bubble. (*b*) The bed section after passage of one bubble.

As more bubbles rise through the bed, so more particles are drawn up to the surface and more flow down, so that the particles are mixed by the bubble-induced, top-to-bottom, particle circulation. If, however, the bubble paths are widely separated laterally, then the bed agitation by the bubbles will be non-uniformly distributed, as shown in figure 2.14. For uniform mixing throughout the bed it is therefore desirable to have the bubble paths close together and uniformly distributed. The close spacing of distributor orifices or the use of porous plate distributors promotes this in the lowest regions of the bed, but bubble coalescence as they rise through the bed, or tendencies for the bubbles to follow preferred paths, lead to lateral non-uniformity of bubbling and, hence, poor lateral mixing of the particles in the presence of good vertical mixing. For further reading the reader is referred to the work of Whitehead (1971), Potter (1971), Whitehead *et al* (1976), Gabor (1971) and Rowe (1971).

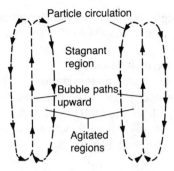

Figure 2.14 A demonstration of poor lateral mixing with widely spaced bubble paths.

The above, then, is the basic mechanism by which particles of Geldart's category B become mixed in the bed. With category A particles, which expand uniformly without bubbling when first fluidized, Rowe (1971) pointed out that such fine powders behave in a slightly different way. In addition to the mechanism outlined above, a bubble passing through category A powders induces a motion corresponding to turbulence in its wake, so that a kind of eddy diffusion assists the mixing process.

Category D particles, usually of a size greater than about 600 mm and/or of high density, are generally characterized by a less effective solids mixing than categories A and B because of the different gas flow and bubble coalescence patterns associated with them.

2.6.2 Mixing and segregation of particles

Within the bubbling regime, the greater the fluidizing velocity, the more violent the bubbling, and the more intense the mixing of particles. However, segregation of particles can occur if there is a sufficient density difference and size difference between the particles, the high-density, larger particles tending to sink to the bottom of the bed. Generally, density differences exert a greater influence on the tendency to segregate than size differences. If it is desired to promote segregation, then the gas velocity should be reduced to a value just greater than the minimum fluidizing velocity. When the density differences are large the denser material will sink to the bottom of even a very highly fluidized bed. For further reading see Rowe and Everett (1972a, b, c).

All this has implications for a wide range of equipment and demonstrates the need for good control of the density and size range of particles in fluidized bed equipment or the provision for removal of jetsam or flotsam. No reliable general predictions of conditions under

50   Particles and Fluidization

which unacceptable segregation occurs can be given; experiments with a representative sample of bed material offer the best guide.

2.6.3 Terminal velocity of particles

If a single particle falls freely under gravity in the atmosphere, it will accelerate until its velocity is such that the drag exerted by the surrounding air is equal to the gravitational force (less the buoyancy force); thus the final steady velocity, known as the 'terminal velocity' U_t, for a sphere of diameter d_s is given by

$$(\rho_p - \rho_f)g\frac{\pi d_s^3}{6} = \tfrac{1}{2}\rho_f U_t^2 S C_D \qquad (2.67)$$

where S is the frontal area of a sphere, $\pi d_s^2/4$, C_D is the drag coefficient (determined experimentally), ρ_p and ρ_f are the densities of solid and gas (air), respectively, and g is the gravitational acceleration.

The drag coefficient C_D depends upon the Reynolds number Re_p, where

$$Re_p = \frac{\rho_f U_t d_s}{\mu_f}. \qquad (2.68)$$

No simple relationship expresses the variation of drag coefficient with Reynolds number. Two relationships often quoted are

$$C_D = \frac{24}{Re_p} \quad \text{for } Re_p < 0.4 \qquad (2.69)$$

and

$$C_D = \frac{10}{Re_p^{1/2}} \quad \text{for } 0.4 < Re_p < 500. \qquad (2.70)$$

Substituting these expressions into (2.67) yields

$$U_t = \frac{(\rho_p - \rho_f)g d_s^2}{18\mu_f} \quad \text{for } Re_p < 0.4 \qquad (2.71)$$

and

$$U_t = \left(\frac{4(\rho_p - \rho_f)^2 g^2}{225\mu_f \rho_f}\right)^{1/3} d_s \quad \text{for } 0.4 < Re_p < 500. \qquad (2.72)$$

However these expressions tend to be inaccurate for non-spherical particles, while their refinement to take account of particle shape is a matter of some controversy. Approaches vary from multiplying the terminal velocity for spherical particles (obtained from the above expressions using the diameter of a sphere, having the same volume as the particle, for d_s) by a correction factor η (see Pettyjohn and Christiansen (1948) and equation (2.73) below). For $Re_t < 0.2$

$$\eta = 0.843 \ln\left(\frac{\varphi}{0.065}\right) \qquad (2.73)$$

where φ is the sphericity (see equation (2.1)).
For $Re_t > 1000$

$$\eta = 5.31 - 4.88\varphi. \qquad (2.74)$$

An alternative method is to use a chart shown in Kunii and Levenspiel (1969), calculated from Brown et al (1950).

A comparison between the various methods used for estimating the terminal velocity U_t can thus lead to differing values. Allowances should be made for this uncertainty and for the variation of gas velocity across the duct cross section (see figure 2.15), when making use of terminal velocity data.

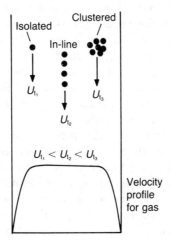

Figure 2.15 The effect of particle grouping on the terminal velocity in a duct with upward flowing gas.

A further important point to realise is that in process plant, frequently, particles are not isolated from each other, sometimes they fall as agglomerates because they stick together (e.g. cohesive powders or sintered fines) or, if the particles are individually separated, they can be shielded by other particles beneath them (see figure 2.15). Such situations give rise to terminal velocities considerably higher than that of a single particle (see Geldart 1981).

The behaviour of particles in a duct having upward flowing gas is further influenced by the fact that the velocity distribution across the duct will be parabolic, if in laminar flow, so that whether a particle or cluster will be entrained in the gas or not depends upon its radial position in the duct. In the case of turbulent flow, random fluctuations of velocity make particle behaviour even more unpredictable. Sagoo (1981) describes such experiences with a falling cloud heat exchanger.

2.6.4 Elutriation

The terminal velocity, U_t, of the smallest particle in the bed is important conceptionally in that it constitutes the maximum gas velocity of the gas at which the fluidized bed should be operated, if entrainment of particles is to be avoided. It has to be said, however, that such a simple concept is confounded in practice by the observation that fluidized beds commonly operate at significantly higher velocities than this without unacceptable rates of particle elutriation from the bed. Elutriation thus has more complex mechanisms than can be explained on the basis of the terminal velocity of a single particle.

Entrainment and elutriation are terms frequently interchanged. The term elutriation when used here will mean the transportation of the finer particles from a wide size range rather than entrainment of the whole size range. The subject of elutriation is not properly understood, yet it is often one of the most important factors limiting the design and the way in which a fluidized bed reactor may be operated.

Particulate solids in fluidized beds are invariably composed of a range of particle sizes and, perhaps, densities, while many solid materials undergo attrition when fluidized, or fragmentation due to thermal stresses when put into a hot fluidized bed. Attrition and fragmentation lead to the production of fines. These fines become entrained in the gas leaving the bed; they may have to be removed from the off-gas stream for environmental reasons, or because of the effect of the loss of bed material on the economics of the plant, or for process reasons. These requirements have to be paid for by the extra costs of gas cleaning equipment and the handling of the elutriated solids.

Most research work on elutriation has been carried out only on small-scale equipment and has led to some empirical correlations whose predictions differ by an order of magnitude or more when applied to conditions different from where they were obtained. The following illustrations may help provide a little insight into this very incomplete field of knowledge.

In a fixed bed, fines can be gradually swept out of the bed by the gas passing through the voids between the large particles. The advent of bubbles in the fluidized state carries particles up with them as they rise through the bed, exchanging particles between the bubble wake and the particulate phase of the bed, as shown in figure 2.16. Upon reaching the free surface the bubble bursts, scattering particles into the above-bed (freeboard) zone. Fortunately, only a very small fraction of the solids

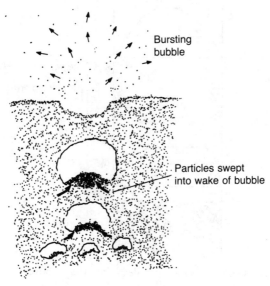

Figure 2.16 Elutriation due to the upward transport of particles by bubbles and the bursting of a bubble.

carried up by the bubble is thrown up into the freeboard zone. However, the upward component of the velocity of particles which are projected into the freeboard may be less than the bubble velocity just before bursting, while other particles may be travelling at about bubble velocity and faster from bubbles coalescing near the bed-free surface. There is thus a spectrum of particle launching velocity v_0, both in magnitude and direction. Those particles whose terminal velocity is small compared with the gas velocity U will be entrained in the gas and elutriated from the system, while those whose terminal velocity is larger than the gas velocity will fall back to the bed surface. The fate of the particles between these extremes depends upon their launch velocity and the gas velocity distribution in the above-bed region. The bursting of bubbles causes turbulence in the freeboard zone, which decays with distance above the bed. This turbulence may assist diffusion of fines towards the slower-moving gas in the boundary layer adjacent to the wall (Pemberton and Davidson 1984), as shown in figure 2.17. Once in the boundary layer the particles tend to fall back to the bed. There is thus a variation in the solids loading (kg m^{-3} of gas or kg kg^{-1} of gas), both laterally and longitudinally, in the above-bed region. The further above the free surface of the bed, the smaller the solids loading. The region above the bed in which the solids loading diminishes is referred to as the 'transport disengagement zone (or height)' and it must be sufficiently high to allow the solids loading to decay to an acceptable value.

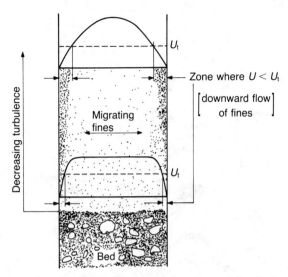

Figure 2.17 Gas velocity profiles in the above-bed region. U_t = terminal velocity of fines.

For a continuously operating fluidized bed, the instantaneous rate of elutriation, E_i, of solids of size d_i can be expressed as

$$E_i = K_i A x_i \tag{2.75}$$

where K_i is the mass of solids transported across a plane parallel to the surface of the bed per unit cross sectional area of bed per unit time, A is the cross sectional area of bed and x_i is the mass fraction of particles of size d_i in the bed.

Clearly, K_i is a function of the height y above the free surface of the bed (indeed, also of lateral position in the duct, in which case A should then be the element dA).

The total rate of elutriation of all sizes of particles will then be

$$E_t = \sum (K_i A x_i). \tag{2.76}$$

The average solids loading of the off-gas, ρ_m, will then be

$$\rho_m = \frac{E_t}{UA} \tag{2.77}$$

where U is the gas velocity at a height y above the bed surface, while the solids loading, ρ_i, due to particles of size d_i only is

$$\rho_i = \frac{E_i}{UA}. \tag{2.78}$$

The value of the rate constant K_i must be known in order to calculate the rate of particle loss. Unfortunately, K_i cannot be calculated theoretically. The reader is referred to Large *et al* (1976), Horio *et al* (1980, 1984), Morooka *et al* (1984), Chan and Knowlton (1984) and Geldart (1986) for more detailed study, but experiments conducted on equipment similar to the plant under consideration give the only quantitative guide to predicting elutriation rates at present.

A reliable model of elutriation applicable to large-scale beds, which can link parameters to the elutriation rates, does not exist at present and is badly needed.

2.6.5 Solids transport

Once particles are fluidized they will flow readily; it has already been seen that a bubbling action leads to the transport and mixing of solids within a fluidized bed, and this has led to the wide use of fluidized beds for carrying out chemical reactions. The fluid-like properties of fluidized solids can also be exploited to transport solids from one location to another, often over long distances, e.g. the hydraulic and pneumatic transport of solids by pumping along a pipeline. In such cases the solids are entrained in the fluid and swept along the pipe by it. When the concentration of particles in the fluid is very dilute compared with that in a bubbling fluidized bed, these are termed 'dilute-phase systems'. No distributor is required in the base of a horizontal pipeline of dilute-phase systems.

Other fluidized solids systems, such as an open channel having a porous base through which fluidizing gas passes into the bed, can be used to make the bulk of the solids flow when the channel is inclined slightly, the amount of bed expansion being relatively small; such systems are therefore termed 'dense-phase systems'. A further type of dense-phase system is one in which slugs of particles can be forced along a pipeline; this is analogous to the slugging regime of a fluidized bed. Figures 2.18–2.20 illustrate such systems.

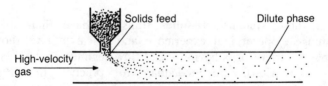

Figure 2.18 Dilute-phase solids transport.

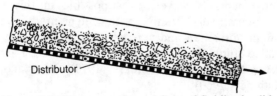

Figure 2.19 The open channel flow of fluidized solids.

56   Particles and Fluidization

Figure 2.20 Slug flow transport.

Other solids flow situations arise when solids have to be transported from one vessel to another, or circulate around a system connected by pipelines, as shown by the downcomers and risers in figure 2.21.

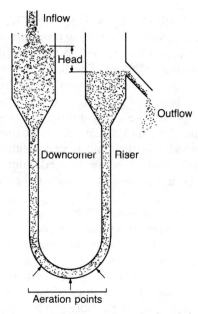

Figure 2.21 Solids flow under a hydrostatic head.

All of these systems of solids transport have their merits and disadvantages; operational experience and research have shown once again that a variety of behaviours and flow regimes exist. Theoretically based design criteria are inadequate for practical design purposes because they provide an insufficient description of what happens in reality; actual designs have to be based on empirical relationships and practical judgement. However, it is possible in this section of the chapter to discuss, very briefly, a few elementary considerations of particle flow in horizontal and vertical ducts and provide references for a more detailed study: Clift (1981), Knowlton (1986), Large et al (1976), Leung (1980), Matsen (1982), Stoess (1970), Wen and Galli (1971), Zenz and Othmer (1960).

Mechanics of two-phase (solid/gas) flow

The normal principles of fluid mechanics can be applied to a mixture of solids and gas moving in a pipeline. Consider first the equation of continuity, referring to figure 2.22. Suppose a compressor or blower delivers a mass flow rate of gas, $\dot{m}_g$, into a pipe of cross sectional area A and solid particles are introduced downstream of the blower at a mass flow rate $\dot{m}_s$.

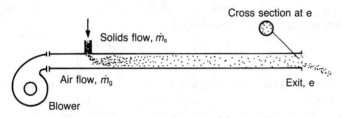

Figure 2.22 Pneumatic transport of solids.

(a) Continuity. If the flow is steady and all particles entering the pipe are delivered at the exit, e, from the pipe, then the mass flow rate of the mixture at e, $\dot{m}_{me}$, is

$$\dot{m}_{me} = \dot{m}_g + \dot{m}_s. \tag{2.79}$$

The solids concentration, i.e. the mass of solids per unit volume of mixture, C_{se}, at the exit is

$$C_{se} = \dot{m}_s / \dot{V}_m. \tag{2.80}$$

Similarly, the concentration of gas at the exit, C_{ge}, is given by

$$C_{ge} = \dot{m}_g / \dot{V}_m \tag{2.81}$$

where $\dot{V}_m$ is the volumetric flow rate of the mixture. If the transport gas is treated as being incompressible, then

$$\dot{m}_g = \rho_g F_g A U_g \tag{2.82}$$

where ρ_g is the gas density, A is the cross sectional area of pipe, F_g is the fraction of cross sectional area of pipe occupied by gas (see the inset of figure 2.22 (note that F_g amounts to 'area voidage')) and U_g is the gas velocity between particles. Similarly, the mass flow rate of solids, $\dot{m}_s$, is

$$\dot{m}_s = \rho_s F_s A U_s \tag{2.83}$$

where ρ_s is the density of the solid particle material, F_s is the fraction of the pipe cross sectional area occupied by solids and U_s is the solids velocity.

Note, that

$$F_s + F_g = 1. \tag{2.84}$$

The volumetric flow rate of the mixture, $\dot{V}_m$, is given by

$$\dot{V}_m = F_g A U_g + F_s A U_s. \tag{2.85}$$

Thus, substituting equations (2.82) and (2.85) into (2.80) gives the average mass concentration per unit volume of mixture, C_{se}, at the exit from the pipe in terms of velocities and fractions of cross sectional area occupied by gas and solids respectively:

$$C_{se} = \frac{\rho_s F_s A U_s}{F_g A U_g + F_s A U_s} = \frac{\rho_s F_s U_s}{F_g U_g + F_s U_s}. \tag{2.86}$$

(Notice that this mass concentration per unit volume of mixture at the exit is different from the mass concentration of solids within the pipe itself. Thus, the mass of solids m_s, enclosed within the pipe of length L and cross sectional area A, is given by

$$m_s = \rho_s A F_s L \tag{2.87}$$

and, hence, the mass concentration C_{sp}, within the pipe

$$C_{sp} = \rho_s F_s. \tag{2.88}$$

Equation (2.88) is not the same as equation (2.86), except when the velocity of the solids and gas are equal; this is an extremely rare occurrence.)

Except for the case when the mixture is flowing downwards through a pipe, the velocity of the solids, U_s, is less than the gas velocity U_g.

Equation (2.86) allows a distinction to be made between 'dilute-phase systems', where F_s is small compared with F_g, and 'dense-phase systems', where F_s is of a similar order to F_g. Thus, a given mass flow rate of solids could leave the pipe at the exit having a high concentration C_{se}, with a low velocity U_s, and F_s being large, i.e. the dense-phase case, or vice versa. The choice of whether to operate as a dilute-phase system or a dense-phase system will be decided initially anyway by the economics of the system, what transport gas supplies are available on the site of the plant, etc. One important element in the running costs of the system is the amount of pumping power required to operate it. The power, $\dot{W}$, required to pump the required mass flow rate of solids, $\dot{m}_s$, along the pipe is given by

$$\dot{W} = (p_1 - p_2)\dot{V}_m \tag{2.89}$$

where p_1 is the pressure at the entry to the pipe and p_2 the pressure at the exit. An equation to estimate the pressure drop across the pipe (which may include bends and changes of section) is therefore needed.

A further point needs to be mentioned here, namely that a transition from the dilute phase to the dense phase or vice versa is not smooth; for example, 'choking' in vertical flow or 'saltation' in horizontal flow, or other phenomena, can occur and it is essential to design so that operation at values of solids concentrations near to which such instabilities arise is avoided.

We proceed now to consider momentum.

(b) *Momentum.* Referring to figure 2.23, we consider the forces on an element of gas/solids mixture having length dx, flowing upwards along a straight pipe of cross sectional periphery P and area A, inclined at an angle θ to the horizontal. The pressure on the upstream side of the element is p and on the downstream side is $(p + dp)$; the shear stress at the wall of the duct is τ.

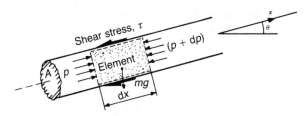

Figure 2.23 Stresses on an element of gas/solid mixture during transport.

The net force in the direction of motion, $A dp$, must overcome the shear stress at the wall, the gravitational component of the weight of the element of mixture and provide an accelerating force:

$$A dp = \tau P dx + [(C_s + C_g) A g \sin \theta] dx + (\dot{m}_s dU_s + \dot{m}_g dU_g)$$
$$\text{(shear)} + \text{(gravity)} + \text{(acceleration)}. \qquad (2.90)$$

For simplicity, consider flow of the two-phase (solids + gas) mixture along a horizontal pipe; as depicted by figure 2.24, then, the pipe may be considered to be made up of two regions, the first being that in which the solids are introduced and accelerated from their inlet velocity U_{si} to their final transport velocity U_s; the gas velocity will also increase in this region to its final velocity U_g between the particles. The dominant term in equation (2.90) in this region is the acceleration term $(\dot{m}_s dU_s + \dot{m}_g dU_g)$. The second region is downstream of this, where the flow is fully developed, so that the only relevant term on the right-hand side of equation (2.90) is the first term $\tau P dx$. Thus, in the acceleration region, integration of the acceleration term alone in equation (2.90) yields

$$\Delta p_{acc} = \frac{1}{A}[\dot{m}_s(U_s - U_{si}) + \dot{m}_g(U_g - U_{gi})]. \tag{2.91}$$

Note, here, that the solids acceleration term, $\dot{m}_s(U_s - U_{si})$, in equation (2.91) will in all practical cases dominate over the gas acceleration term.

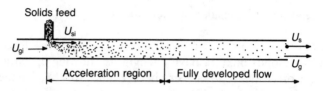

Figure 2.24 The development of two-phase flow in pneumatic transport.

Integration of the shear stress term, $\tau P dx$, alone within the fully developed flow region gives

$$\Delta p_{fd} = \frac{\tau P L_{fd}}{A} \tag{2.92}$$

and, hence, the overall pressure drop Δp is

$$\Delta p = \Delta p_{acc} + \Delta p_{fd}. \tag{2.93}$$

An estimate of the length of pipe required to accelerate the solids to their fully developed flow condition, the shear stress at the pipe wall and the influence of bends in the pipe is beyond the intended scope here.

Instead it is timely to mention, very briefly, some important considerations of solids transport along a horizontal pipe.

(c) Saltation velocity. The system shown in figure 2.24 is a horizontal pipeline; it has distinct regimes within which it functions satisfactorily. If the solids flow rate is kept constant and the gas flow velocity is maintained at a high value, then the solid particles will be entrained in the gas and be discharged at the end of the pipe. The gradual reduction of gas velocity will eventually result in particles starting to fall out of suspension and settling on the bottom of the pipe where, although they can still be transported, the regime has changed to an unstable one (see, for example, Wen and Galli 1971). The gas velocity at which this occurs is called the 'saltation velocity', U_{salt}. Methods of estimating saltation velocity have been devised (see Zenz 1964, Duckworth 1976); they are based upon experimental observations and are commonly used for design. However, if practicable, it is safer and also instructive to carry out the experiments to determine saltation with the actual particles and

conveying gas proposed because of the dependence of the performance of the system upon particle size, size range, shape, moisture content and hygroscopy, as well as the properties and humidity of the conveying gas and the arrangement of the conveying pipeline. Any further reduction in gas flow below the saltation velocity will eventually lead to intermittent flow of gas and solids and, finally, plugging of the pipe.

The type of experiment that can be used to determine saltation velocity is illustrated by figure 2.25, which is taken from the work of Zenz (1957) and Zenz and Othmer (1960). Curve AB represents the pressure drop/gas velocity relationship when no solids are in the pipe. Curve CD represents pressure drop versus gas velocity at a fixed rate of solids flow $\dot{m}_s$ in the fully dispersed dilute phase. When the gas velocity is reduced to the saltation velocity at point D, there is a discontinuity and an abrupt rise in pressure drop along the pipe to some value E as the solids settle to the bottom of the pipe. A continued reduction in gas velocity causes still more solids to settle out, partially blocking the pipe, so that the pressure drop rises further until F is reached, where the pipe has become blocked with solids, and the gas can only percolate through as in a packed bed. Lines GHKL and MNPQ show the relationship at

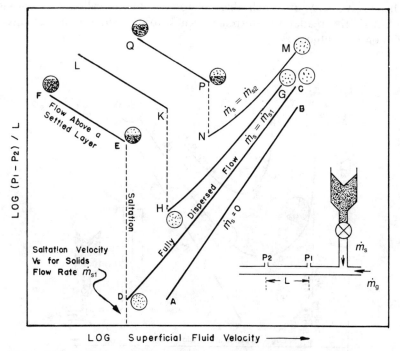

Figure 2.25 A schematic representation of fluid–solids flow and characteristics in horizontal transport. (After Zenz and Othmer 1960.)

62 Particles and Fluidization

progressively higher solids flow rate. The saltation velocity is therefore a function of the solids concentration. In dilute-phase conveying systems, values of 'solids loading', i.e. the mass of solids per unit mass of conveying air can lie between 0.1 and 5 (Wen and Galli 1971).

A discussion of dense-phase conveying will be left to a later chapter.

(d) Vertical transport of solids. Figure 2.26 shows the pressure drop versus gas velocity relationship in a vertically upward flow of gas and solids, taken from Zenz and Othmer (1960). Curve AB shows the relationship when no solids are fed into the pipe. When the flow rate of solids is fixed at same value $\dot{m}_s$ we refer to curve CDE. As the gas flow is reduced from that at C, where the particles are fully dispersed throughout the gas, the pressure drop decreases, the solids concentration increases and the particle velocity reduces. At point D the solids concentration (or density of the mixture) has increased to the point at which the static pressure head begins to exceed the drag force exerted by the gas on the particles. A further reduction of gas velocity causes a further increase in solids concentration and an increase in pressure drop until at point E the gas drag can no longer support the suspension of solids, whence they collapse and the system operates in a slugging flow. The gas velocity at which this collapse occurs is called the 'choking velocity'; the reader is referred to Knowlton (1986) for further discussion on this subject.

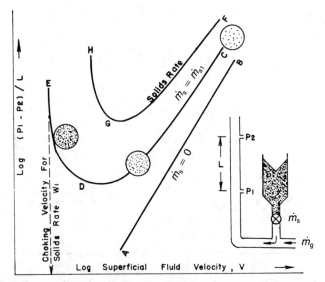

Figure 2.26 A schematic representation of flow characteristics in vertical transport. (After Zenz and Othmer 1960.)

Examples for Chapters 1 and 2

1. A system is required for bringing a continuous flow of gas into good contact with the surface of granules so that heat and mass transfer can occur at the solid–gas interface. Assuming that the granules can be fluidized readily if required and that they are to be processed a batch at a time, make sketches to show some alternative proposals for achieving this objective and make a very brief qualitative assessment of the advantages and disadvantages of each alternative.

2. Silica sand particles of bulk density 1600 kg m^{-3} are loosely packed to a depth of 0.3 m in a small vessel of 1.7 m diameter. The vessel has a porous base, or distributor, through which air can flow into the bed. Determine the pressure drop across the bed when incipiently fluidized.
   The pressure drop (in N m^{-2}) across the distributor at ambient temperature and pressure is related to the superficial air velocity U (in m s^{-1}) by the equation

$$\Delta p_d = 0.13U + 0.03U^2.$$

   Determine the pressure drop across the bed and distributor when fluidized at ambient temperature and pressure, where the minimum fluidizing velocity of the particular sand particles is 0.19 m s^{-1}.
   Also determine the pumping power absorbed by the system when fluidized with ambient air at 0.5 m s^{-1}.

3. If, with the bed of particles described in Example 2, the air is supplied to the bed at only half the minimum fluidizing velocity, determine the pressure drop across the bed, across the distributor, and overall; also calculate the pumping power absorbed by the bed and distributor combined.

4. The operating conditions with the bed of particles described in Example 2 are likely to be altered by the fluidizing air being supplied at a higher temperature. If the velocity of air was to remain the same, how would you expect the pressure drops
   (a) across the bed and
   (b) across the distributor,
   when the bed is incipiently fluidized, to be affected?
   (Hint: refer to tables of the density and viscosity of air.)

5. A sieve analysis of a sample of spherical particles is made up as follows:

Mean aperture size (μm)	275	428	550	655	780
Mass of particles (g)	30	90	120	70	40

Estimate the mean particle size, the nominal size range, and the relative size range, based upon the fraction of the sample lying between mass fractions 0.16 and 0.84.

Also estimate the total surface area of particles in the bed.

6. The sieve analysis shown in Example 5 was found on later examination to be in error in that the mass of particles of mean size 275 μm was 65 g instead of 30 g. What effect does this have on the estimated value of the mean particle size, the relative size range and the total surface area of particles?

7. The shape of a single particle of powder was found on microscopic examination to be almost cubic. Estimate the surface area : volume ratio, the sphericity and the size of the particle according to two alternative definitions. What size, in terms of the length of the side of a cube might be indicated by sieving such particles using sieves having square shaped apertures?

8. A sample of particulates has a mass of 980 g and is made up of a wide range of particle sizes and shapes. The sample is poured gently into a calibrated volume measuring vessel and it is observed that the free surface coincides with the 650 ml mark. Determine the bulk density of the particles.

The cylinder is then vibrated and it is found that the free surface sinks by 50 mm. Explain why this occurs.

A representative sample of the particulates is then placed in a specific gravity bottle of volume 25 ml. The mass of the particles in the bottle is 35.3 g. A low-viscosity liquid of density 680 kg m^{-3} is added to the bottle and the mixture agitated until no further liquid can be accepted; the mass of added liquid is 6.42 g. Determine the particle density and hence the voidage of the vibrated particulates in the measuring cylinder.

9. Estimate the minimum fluidizing velocity using the pressure drop and air flow data for the particles given in the table below. The particles are contained in a vessel of 150 mm diameter.

Air flow rate (l min^{-1})	25	51	77	98	127	150	176	205	225	255	295
Pressure drop across bed (mm water)	31	62	93	124	150	172	183	186	188	188	188

The above data were obtained when the fluidizing air was at a pressure of 1 atm, and a temperature of 27 °C, where the density and viscosity were 1.177 kg m^{-3} and 1.846 × 10^{-5} kg m^{-1} s^{-1}, respectively. The particle mean size by sieve analysis was 520 μm and the particle density was 2610 kg m^{-3}, while the bulk density of a loosely packed bed of these particles was 1490 kg m^{-3}. Estimate the sphericity of these particles.

It is proposed to use these particles in a reactor fluidized by a gas whose properties at operating conditions of 1.3 atm and 177 °C are a density of 1.55 kg m^{-3} and a viscosity of 22.0 × 10^{-5} kg m^{-1} s^{-1}. Estimate the minimum fluidizing velocity when fluidized by this gas

(i) using the above test data and the Ergun equation (2.45),
(ii) using the empirical correlation due to Wen and Yu (equation (2.47)) and
(iii) using the Goroshko equation (2.56) and its terms defined by equations (2.52) and (2.53).

Verify that the Archimedes and Reynolds Numbers are within the range of validity for such operating conditions.

Comment on the different values obtained from the estimates of minimum fluidizing velocity.

10. The minimum fluidizing velocity of a bed of particles is 0.16 m s^{-1} and the bed is to operate at 2.5 times this value. Visual observation suggests that the largest bubbles formed are of 10 mm diameter. Estimate the residence time of a single, isolated bubble if the bed depth when fluidized is 650 mm.

Decide which category of bubble this is and make a sketch of the likely gas flow pattern into and out of the bubble.

11. A bed of particles when loosely packed is 500 mm deep and its voidage is 0.41. When fluidized at 2.1 times the minimum fluidizing velocity of 0.045 m s^{-1}, the bed expands to a depth of 610 mm. Estimate the mean bubble size, velocity, and frequency.

12. A static pressure probe immersed in a bed being fluidized at three times the minimum fluidizing velocity indicates a non-linear variation of static pressure with bed depth, as shown in the table below. The bulk density of the bed when loosely packed is 980 kg m^{-3} and the particle density is 1640 kg m^{-3}.

Height above distributor (m)	0.1	0.2	0.3	0.4	0.5
Static pressure (m water)	0.315	0.258	0.197	0.133	0.0672

A separate test shows that the minimum fluidizing velocity is 0.051 m s^{-1}. Estimate the mean voidage, bubble size, velocity, and frequency at the levels 0.15, 0.25, 0.35 and 0.45 m above the distributor. To what do you attribute the change in apparent mean bubble size with distance above the distributor?

13. In a particular design of reactor, the particle size and fluidizing velocity are fixed, the only variable which can be controlled is bubble size. It is desired to allow the bed to expand as much as practicable. Should the designer try to promote small bubbles or large ones? Give reasons for your answer.

14. A horizontal pneumatic pipeline is 45 m long. Solids are fed into it at negligible velocity and are accelerated by the air passing through it.

A simple test in dilute-phase transport shows that 0.176 m^3 of air at ambient conditions are required to transport 1 kg of solids and that the required air velocity on an empty pipe basis is 32 m s^{-1}. Estimate the pipe diameter required to convey 1000 kg h^{-1} of solids.

If the fraction of the pipe cross sectional area occupied by solids, F_s, is 0.005, estimate the particle and gas velocities, U and U_g, the slip velocity, $(U_g - U_s)$, and the pressure drop required to accelerate the particles to their final velocity.

How would you expect the pressure drop required to accelerate the particles to be affected by a change of slip velocity? Verify by calculation at values of F_s greater than and less than 0.005.

How would you expect the pressure drop where the flow is fully developed to depend upon slip velocity?

References

Abrahamsen A R and Geldart D 1980 Behaviour of gas-fluidized beds of fine powders—part I. Homogeneous expansion *Powder Technol.* **26** 35–46
Allen T 1981 *Particle Size Measurement* 3rd edn (London–Chapman and Hall)
Baeyens J 1973 PhD Thesis University of Bradford
Baeyens J and Geldart D 1973 in *Fluidisation et ses Applications* (Toulouse: Societé Chimie Industrielle) p263
Botterill J S M 1975 *Fluid Bed Heat Transfer* (London: Academic) p7
—— 1983 in *Fluidized Beds: Combustion and Applications* ed. J R Howard (London: Applied Science) p7
Botterill J S M, Teoman Y and Yüriger K R 1982 The effect of operating temperature on the velocity of minimum fluidization bed voidage and general behaviour *Powder Technol.* **31** 101–10
Brown G G et al 1950 *Unit Operations* (New York: Wiley)
Chan I H and Knowlton T M 1984 in *Fluidization* ed. D Kunii and R Toei (New York: Engineering Foundation) pp283–90
Clift R 1981 in *Gas Fluidization* ed. D Geldart (Rugby: University of Bradford–Institution of Chemical Engineers) §0, unpublished

Cranfield R R and Geldart D 1974 Large particle fluidization *Chem. Eng. Sci.* **29** 935-47

Davidson J F 1961 Symposium on Fluidization – discussion *Trans. Inst. Chem. Eng.* **39** 230-2

Davidson J F and Harrison D 1963 *Fluidized Particles* (Cambridge: CUP) p24

—— (ed.) 1971 *Fluidization* (London: Academic)

Davies R M and Taylor G 1950 The mechanics of large bubbles rising through extended liquids and through liquids in tubes *Proc. R. Soc.* A **200** 375-90

de Groot J H 1967 in *Int. Symp. Fluidization* ed. A H H Drinkenberg (Amsterdam: Netherlands University Press) p348

Duckworth R A 1976 Pneumotransport 3 *British Hydromechanics Research Association* paper S5-35

Ergun S 1952 Fluid flow through packed columns *Chem. Eng. Prog.* **48** 89-94

Ergun S and Orning A A 1949 Fluid flow through randomly packed columns and fluidized beds *Ind. Eng. Chem.* **41** 1179-84

Gabor J D 1971 Boundary effects on a bubble rising in a finite two-dimensional fluidized bed *Chem. Eng. Sci.* **26** 1247-57

Geldart D 1972 The effect of particle size and size distribution on the behaviour of gas-fluidized beds *Powder Technol.* **6** 201-15

—— 1973 Types of gas fluidization *Powder Technol.* **7** 285-92

—— (ed.) 1981 *Gas Fluidization* (Rugby: University of Bradford – Institution of Chemical Engineers) unpublished

—— (ed.) 1986 *Gas Fluidization Technology* (Chichester: Wiley–Interscience) pp3-4

Geldart D and Abrahamsen A R 1981 Fluidization of fine porous powders *AIChE Symp. Ser. No* 205 **77**

Glicksman L R, Lord W K and Sakagami M 1985 private communication

Goroshko V D, Rosenbaum R B and Todes O H 1958 *Izv. Vuzor, Neft'i Gaza* **I** 125

Grace J R 1986 Contacting modes and behaviour classifications of gas-solid and other two-phase suspensions *Can. J. Chem. Eng.* **64** 353-63

Horio M, Shibata T and Muchi I 1984 in *Fluidization* ed. D Kunii and R Toei (New York: Engineering Foundation) pp307-14

Horio M, Taki A, Hsieh Y S and Muchi I 1980 in *Fluidization* ed. J R Grace and J M Matsen (London: Plenum) pp509-18

Hovmand S and Davidson J F 1971 in *Fluidization* ed. J F Davidson and D Harrison (London: Academic) ch 5

Howard J R (ed.) 1983 *Fluidized Beds: Combustion and Applications* (London: Applied Science)

Knowlton T M 1986 in *Gas Fluidization Technology* ed. D Geldart (Chichester: Wiley–Interscience) ch 12

Konrad K, Harrison D, Nedderman R M and Davidson J F 1980 Pneumotransport 5 *British Hydromechanics Research Association* Paper E1

Kunii D and Levenspiel O 1969 *Fluidization Engineering* (New York: Wiley) p77, figure 8

Kunii D and Toei R (ed.) 1984 *Fluidization* (New York: Engineering Foundation)

Large J F, Martinie Y and Bergougnou M A 1976 *Int. Bulk Handling Conf.*, Chicago
Leung L S 1980 in *Fluidization* ed. J R Grace and J Matsen (London: Plenum) pp25–68
Matsen J M 1982 Mechanisms of choking and entrainment *Powder Technol.* **32** 21
Moroney M J 1960 *Facts from Figures* 3rd edn (London: Penguin Books) p58
Morooka S, Kago T and Kato V 1984 in *Fluidization* ed. D Kunii and R Toei (New York: Engineering Foundation) pp291–8
Pemberton S and Davidson J F 1984 in *Fluidization* ed. D Kunii and R Toei (New York: Engineering Foundation) pp275–82
Pettyjohn E S and Christiansen E B 1948 Effect of particle shape on free-settling rates of isometric particles *Chem. Eng. Prog.* **44** 157–72
Potter O E 1971 in *Fluidization* ed. J F Davidson and D Harrison (London: Academic) ch 7
Rippin D W T 1959 *PhD Dissertation* University of Cambridge
Rowe P N 1964 Gas-solid reaction in a fluidized bed *Chem. Eng. Prog.* **60** 75–80
—— 1971 in *Fluidization* ed. J F Davidson and D Harrison (London: Academic) ch 4
Rowe P N and Everett D J 1972a Fluidized bed bubbles viewed by x-rays. Part I. Experimental details and the interaction of bubbles with solid surfaces *Trans. Inst. Chem. Eng.* **50** 42–8
—— 1972b Fluidized bed bubbles viewed by x-rays. Part II. The transition from two to three dimensions of undisturbed bubbles *Trans. Inst. Chem. Eng.* **50** 49–54
—— 1972c Fluidized bed bubbles viewed by x-rays. Part III. Bubble size and number when unrestrained three-dimensional growth occurs *Trans. Inst. Chem. Eng.* **50** 55–60
Rowe P N, Foscolo P U, Hoffman A C and Yates J G 1984 in *Fluidization* ed. D Kunii and R Toei (New York: Engineering Foundation) pp53–60
Sagoo M S 1981 The development of a falling cloud heat exchanger – air and particle flow and heat transfer *J. Heat Recov. Syst.* **1** 133–8
Scheidegger A E 1974 *The Physics of Flow Through Porous Media* 3rd edn (Toronto: University of Toronto Press) pp165–7, 33–6
Stergiou L and Laguerie C 1984 in *Fluidization* ed. D Kunii and R Toei (New York: Engineering Foundation) pp557–64
Stoess J A Jr *Pneumatic Conveying* (New York: Wiley–Interscience)
Wen C Y and Galli A F 1971 in *Fluidization* ed. J F Davidson and D Harrison (London: Academic) pp677–708
Wen C Y and Yu Y H 1966 A generalized method for predicting the minimum fluidizing velocity *AIChE J.* **12** 610–12
Whitehead A B 1971 in *Fluidization* ed. J F Davidson and D Harrison (London: Academic) ch 19
Whitehead A B, Gartside G and Dent D C 1976 Fluidization in large gas-solid systems. Part III. The effect of bed depth and fluidizing velocity on solids circulation patterns *Powder Technol.* **14** 61–70

Yates J G 1983 *Fundamentals of Fluidized-bed Chemical Processes* (London: Butterworths) pp15–34
Zenz F A 1957 Minimum velocity for catalyst flow *Petrol. Refiner* **36** 133–42
—— 1964 Conveyability of materials of mixed particle size *Ind. Eng. Chem. Fund.* **3** 65–75
Zenz F A and Othmer D F 1960 *Fluidization and Fluid Particle Systems* (New York: Reinhold)

Bibliography

Hikita T, Ideka M and Asano H 1984 in *Fluidization* ed. D Kunii and R Toei (New York: Engineering Foundation) pp219–24
Matsen J M 1984 in *Fluidization* ed. D Kunii and R Toei (New York: Engineering Foundation) pp225–32
McGuigan S 1974 *PhD Thesis* University of Aston in Birmingham
Pinchbeck P H and Popper F 1956 Critical and terminal velocities in fluidizatioin *Chem. Eng. Sci.* **6** 57–64
Pugh R 1975 *PhD Thesis* University of Aston in Birmingham

3 Fluidized Bed Heat Transfer

3.1 Modes of Heat Transfer

3.1.1 General remarks
All three modes (conduction, convection and radiation) of heat transfer are relevant to processes associated with beds of particles. The relative importance of each mode depends upon circumstances; but, in general, the rate of heat transfer between a bed of particles and surfaces in contact with it is governed by the behaviour of the bed. It might, however, be helpful here to remind the reader of the physical relationships which describe each mode.

3.1.2 Conduction
The heat flux q_{cond} (i.e. the rate of heat transfer per unit cross sectional area normal to the direction of heat flow) is given by the well known Fourier equation

$$q_{cond} = -k\frac{dT}{dx} \tag{3.1}$$

where k is the thermal conductivity of the material and dT/dx is the temperature gradient in the direction of heat flow, x. In the case where a particle is brought into contact with a surface at a different temperature, as shown in figure 3.1, equation (3.1), although fundamental, is insufficient to describe the situation mathematically. A heat diffusion equation for more than one coordinate direction, with appropriate boundary and initial conditions, is required because of the varying nature of the conduction path. Details pertinent to fluidized beds can be found in Botterill (1975).

Equation (3.1) also holds for the conduction of heat through liquids and gases, provided that convective motion is suppressed.

3.1.3 Convection
Heat transfer by convection occurs only with liquids or gases and is dependent upon the movement of the elements of the fluid. Convective

Modes of Heat Transfer 71

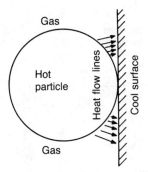

Figure 3.1 Heat flow when a hot particle is in contact with a cool surface—most heat is conducted through the gas film.

motion of the elements can be either 'free', brought about by density differences within the liquid or gas, or 'forced'. In either case, as shown in figure 3.2, an element of fluid is brought into contact with a surface or other element at a different temperature for a short period of time, heat is exchanged between them and the elements interchange continuously. The equation for convective heat flux, $\dot{q}_{conv}$, is given by

$$\dot{q}_{conv} = h(T_f - T_s) \tag{3.2}$$

where h is a heat transfer coefficient, T_f is the fluid bulk temperature (see figure 3.3) and T_s is the surface temperature. The value of the heat transfer coefficient depends upon the rate of replacement of the fluid elements at the surface. Frequent replacement leads to a greater heat

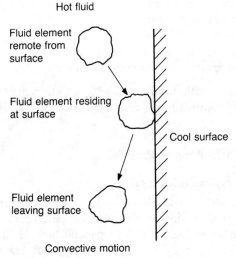

Figure 3.2 The mechanism for convective heat transfer.

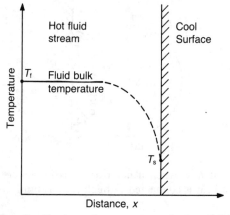

Figure 3.3 The distribution of temperature in hot fluid exchanging heat with a cool wall.

transfer rate than when elements are allowed to reside at the surface for long periods; accordingly, a larger value for the heat transfer coefficient h is found with frequent interchange.

3.1.4 Radiation
Heat can be transferred between two surfaces at different temperatures without the necessity of any physical substance between them. The radiative heat flux $\dot{q}_{rad}$ is given by

$$\dot{q}_{rad} = \varepsilon\sigma(T_1^4 - T_2^4) \tag{3.3}$$

where ε is an exchange factor dependent on surface emissivities and geometrical arrangement
σ is the Stefan–Boltzmann constant
T_1 is the temperature of the hotter surface
T_2 the temperature of the cooler surface.

Generally speaking, with fluidized beds, radiative heat transfer is not significant at temperatures lower than about 600 °C (see Botterill 1983).

3.2 Heat Transfer in Beds of Particles

3.2.1 Gas-to-particle heat transfer
It is perhaps simplest to consider a bed of particles initially at uniform temperature, through which a warmer gas percolates slowly without disturbing the particles, i.e. a fixed or packed bed. Heat will flow from the gas to the particles by convection, the heat transfer coefficient depending upon the flow regime (laminar, transition or turbulent) in the

voids surrounding the particles. However, the overwhelming reason for the highly effective exchange of heat between the gas and the particles is because of the extremely large surface area of particles exposed to the gas. This surface area actually contacted by the gas may only be a fraction of the total surface area of the particles in the bed because the gas may flow through preferred paths among the particles. Thus the gas-to-particle-surface heat transfer coefficient based upon the total surface area of the particles may still be quite small. As the gas velocity through the voids increases, so the heat transfer coefficient at the gas/particle interface will rise, but it is difficult to determine such heat transfer coefficients with any great accuracy. Kunii and Levenspiel (1969) have collected data on gas-to-particle heat transfer in fixed beds over a wide range of size and type of particles and fluid velocity, when air or water was used as the fluid. The data were plotted as non-dimensional groups, Nusselt number Nu ($= hd_p/k_f$) versus particle Reynolds number Re ($= \rho_f U d_p/\mu_f$). This showed that a correlation of the form $Nu = C \times Re^n$, where C and n are empirical constants, is of a limited range of application because of the wide variation in C and n between the various beds of particles. When trying to decide upon suitable values of gas-to-particle heat transfer coefficients in fixed beds, therefore, experiments with the actual particles to be used in a bed under actual operating conditions are to be preferred. This matter will not be pursued farther here.

The other very important point concerning gas-to-particle heat transfer is the fact that the thermal capacity, (heat stored per unit volume), of the gas is very small compared with that of the particles, e.g. the thermal capacity of air at atmospheric pressure is of the order of one thousandth of that of the particles. Thus, on exposure to the particles, the gas temperature would be expected to change to a value close to that of the particles; not the converse. Accordingly, the temperature of the particles dictates the gas temperature in the voids and, with it, the gas density and viscosity; thus the convective gas-to-particle heat transfer coefficient and, indeed, the entire behaviour of the bed, whether fixed or fluidized, are conditioned by particle temperature.

Consider now gas-to-particle heat transfer when the bed is *fluidized*. The amount of data in the literature is small, while the conditions under which they were obtained and the assumptions about flow and gas/particle mixing vary widely. Kunii and Levenspiel (1969) show a plot of data based on the assumption that flow through the bed is plug flow, i.e. that there is negligible transverse variation in gas temperature only in the direction of flow. The resulting correlation was

$$Nu_{gp} = 0.03 \, Re^{1.3} \qquad (3.4)$$

where

74 Fluidized Bed Heat Transfer

Nu = gas-to-particle Nusselt number $(h_{gp}d_p/k_f)$ (3.5)

Re = particle Reynolds number $(\rho_f U d_p/\mu_f)$. (3.6)

It may be instructive to make an estimate of how far into a fluidized bed of particles the fluidizing gas has to penetrate before its temperature difference from that of the particles has fallen to about 10% of that at entry to the bed.

Example 3.1. A bed of particles of mean size 427 μm is fluidized uniformly with air. The air enters the bed at atmospheric pressure at a temperature of 122 °C. The particles are well agitated and maintained at a uniform temperature of 22 °C; at this temperature the fluidizing air velocity is 0.28 m s^{-1} and the observed bed voidage is 0.42. Using the data in table 3.1 for air properties at a mean air temperature of 77 °C, estimate how far the air has to penetrate into the bed for its temperature to become 32 °C.

Table 3.1 Properties of air at atmospheric pressure and 77 °C.

Density	1.009 kg m^{-3}
Thermal conductivity	3.003 × 10^{-5} kW m^{-1} K^{-1}
Viscosity	2.075 × 10^{-5} kg m^{-1} s^{-1}
Specific heat at constant pressure	1.008 kJ kg^{-1} K^{-1}

First calculate the air velocity U at the mean temperature of 77 °C:

$$U = 0.28 \times (77 + 273)/(22 + 273) = 0.332 \text{ m s}^{-1}.$$

Then calculate the particle Reynolds number at this air velocity:

$$Re = \frac{1.009 \times 0.332 \times 427 \times 10^{-6}}{2.075 \times 10^{-5}}$$

$$= 6.89.$$

From equation (3.4), calculate the Nusselt number

$$Nu = 0.03 \times (6.89)^{1.3}$$

$$= 0.369.$$

The gas-to-particle heat transfer coefficient h_{gp} is obtained from the definition of the gas-to-particle Nusselt number, $Nu_{gp} = h_{gp}d_p/k_f$, so that

$$h_{gp} = \frac{Nu_{gp}k_f}{d_p} \quad (3.7)$$

$$= \frac{0.369 \times 3.003 \times 10^{-5}}{427 \times 10^{-6}}$$

$$= 0.0260 \text{ kW m}^{-2} \text{ K}^{-1}.$$

Referring to figure 3.4 which shows the air and particle temperature distribution in the region immediately above the distributor, we calculate the value of y as follows. The mean temperature difference between the particles and fluidizing air will be taken as the logarithmic mean temperature difference (LMTD), which is normally used in heat exchanger calculations. This is given by

$$\text{LMTD} = \frac{\theta_1 - \theta_2}{\ln(\theta_1/\theta_2)} \quad (3.8)$$

where θ_1 and θ_2 are the temperature differences between the air and bed at entry and at the penetration distance y, respectively.

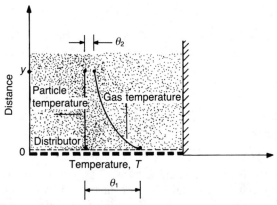

Figure 3.4 The air and particle temperature distribution near the distributor of a fluidized bed.

The rate of heat transfer, $\dot{Q}$, from the air to the particles over the distance y is then given by

$$\dot{Q} = h_{gp} A_p (\text{LMTD}) \quad (3.9)$$

where A_p is the surface area of the particles. The rate of enthalpy loss by the air, Q, is given by

$$Q = \dot{m} C_p (\Delta T) \quad (3.10)$$

where $\dot{m}$ is the mass flow rate and C_p is the specific heat at constant pressure. Hence, equating equations (3.9) and (3.10) and noting by inspection of figure 3.4 that the numerator of equation (3.8), $\theta_1 - \theta_2$, is the temperature decrease, ΔT, of the fluidizing air (since in this case, the temperature of the particles is constant), the following expression for the particle surface area, A_p, is obtained:

$$A_p = \frac{\dot{m} C_p \ln(\theta_1/\theta_2)}{h_{gp}}. \quad (3.11)$$

Considering a unit planform area of distributor, the mass flow rate m is simply the 'mass velocity' $\rho_f U$, so that equation (3.11) becomes

$$A_p = \frac{\rho_f U C_p \ln(\theta_1/\theta_2)}{h_{gp}}. \tag{3.12}$$

Also

$$A_p = \text{number of particles} \times \pi d_p^2 \tag{3.13}$$

while the volume of particles, V_p, is

$$V_p = \text{number of particles} \times (\pi d_p^3/6) \tag{3.14}$$

and, since for a unit planorm area of bed

$$V_p = y(1 - \varepsilon) \tag{3.15}$$

where ε is the bed voidage, the distance y into the bed, to which the air has to penetrate for its temperature to be lowered to 32 °C, is

$$\begin{aligned} y &= \frac{\rho_f U C_p d_p \ln(\theta_1/\theta_2)}{6 h_{gp}(1 - \varepsilon)} \\ &= \frac{1.009 \times 0.332 \times 1.008 \times (427 \times 10^{-6}) \ln(100/10)}{6 \times 0.0260 \times (1 - 0.42)} \\ &= 3.68 \times 10^{-3} \text{ m, i.e. 3.68 mm}. \end{aligned} \tag{3.16}$$

Thus, the distance the fluidizing air has to penetrate for its temperature to be close to that of the main bed of particles is quite small, less than 10 particle diameters in this case. If the air is not uniformly distributed, so that the region contains de-fluidized zones, or some of the particles have agglomerated, then the above calculation will lack validity.

The simple correlation equation (3.4) was used to determine the gas-to-particle heat transfer coefficient in the above example. A more recent correlation, which takes into account the depth of the bed, is due to Kato *et al* (1979) who carried out experiments on the drying rate of particles of activated alumina, of sizes ranging from 324 to 558 μm, in relatively shallow packed fluidized beds (15–50 mm deep), the packing being open-ended cylindrical screen wire nets. This investigation yielded a correlation for the gas-to-particle heat transfer which indicated that the Nusselt number depended upon bed depth as well as the particle Reynolds number. The correlation was applicable over the range of particle Reynolds number $3 < Re < 50$, and is given by

$$Nu_{gp} = 0.59 Re^{1.1} (d_p/L_f)^{0.9} \tag{3.17}$$

where L_f is the bed depth.

The tests showed that the influence of packing size was small. If equation (3.17) is to be used in the above example, instead of (3.4), then the bed depth L_f would have to be given. Suppose that $L_f = 50$ mm, then the corresponding distance y the air would have to penetrate for its temperature to become 32 °C would be 20 mm; but with a bed depth of $L_f = 15$ mm, y would be 6.8 mm. Such variations in predictions show that it is very important to carry out tests rather than rely on correlations alone.

3.2.2 Bed-to-surface heat transfer
Heat will be transferred between a bed of particles, its containing surfaces and any surface immersed within the bed if a temperature difference exists between them.

The heat transfer between the bed and surface in contact with it will be made up of three components, which are regarded as being additive.

(i) That due to particles being brought into contact with the surface, residing there and then being replaced by fresh particles. This component and the corresponding heat transfer coefficient h_{pc} are termed the 'particle convective' terms.

(ii) Convective heat transfer between the interphase gas and surface. This component and the corresponding heat transfer coefficient h_{gc} are termed the 'interphase gas convective' terms.

(iii) A radiative component of heat transfer, with a corresponding heat transfer coefficient h_{rad}.

Thus, the bed-to-surface heat transfer coefficient h_{bs} is given by

$$h_{bs} = h_{pc} + h_{gc} + h_{rad}. \tag{3.18}$$

The relative magnitude of these components and their controlling parameters depends upon the particular particles, gas properties, bed operating regime and behaviour; see Botterill (1983) for a detailed discussion. It will be sufficient here, however, to make brief illustrative comments about each component.

3.2.3 Particle convective component, h_{pc}
For simplicity, consider figures 3.5(a) and (b) which illustrate the situation when a single, hot particle is brought into contact with a continuously cooled surface for a short period of time and then removed, the cycle being repeated continuously with fresh particles. When the particle is in contact with the surface, actual solid-to-solid contact only occurs over a microscopically small area (see Decker and Glicksman 1981) so that the main path through which heat is transferred from the particle to the surface is via the gas separating the particle and the surface. Once the particle is removed, heat transfer from it ceases until a new particle arrives at the surface.

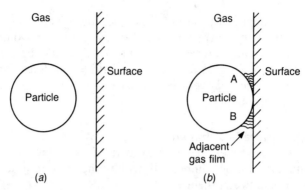

Figure 3.5 A single hot particle being brought into contact with a cool surface. (*a*) The particle remote from the surface—negligible heat transfer. (*b*) The particle residing at the surface—most heat flows through the gas film adjacent to the sector AB.

The parameters governing the amount of heat received by the surface are as follows.

(i) Particle residence time t_p and frequency of contact f_c. The shorter the residence time and the more frequently the particles contact the surface, the greater the time-average temperature difference ΔT between the particles and the surface. This generally controls the achievable heat transfer rate.

(ii) Thermal conductivity of the gas. The larger the thermal conductivity, the more readily heat will flow. Because thermal conductivities of gases are generally low compared with those of solids, this is the limiting factor in the heat transfer process.

(iii) Specific heat capacity of the particle. The larger the specific heat, the less the particle temperature will fall during its residence period and, hence, the larger the time-average temperature difference between the particle and the surface.

(iv) Size of particles and the number in contact per unit area of surface. The greater the fraction of the surface which is exposed to particle contact, the greater the heat received by the surface. Further, if the particles arrive and depart from the surface as clusters or 'packets', as shown in figure 3.6, the smaller particles provide a greater density of contact and a shorter limiting conduction heat transfer path through the gas film adjacent to the surface.

Thus, to promote heat transfer between a bed of particles and a surface, the densest packing of the smallest particles having the shortest residence times and most frequent replacement is required, together with the most conductive gas.

Heat Transfer in Beds of Particles 79

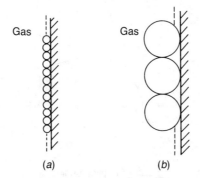

Figure 3.6 Illustrating the effect of particle size on the number of contact points and the thickness of gas film between particles and the surface. (a) Small particles with a large number of points of contact. A large fraction of the particle surface is close to the plane surface. (b) Large particles with a small number of points of contact. A smaller fraction of the particle surface is close to the plane surface.

Several mathematical models describing the particle convective heat transfer process have been proposed, but, as Botterill (1975, 1983) points out, they cannot be used generally because conditions adjacent to the heat transfer surface and values of important parameters, particularly particle residence times and contact frequency, are not known.

Figure 3.7 shows the relationship between the bed-to-surface heat transfer coefficient and the velocity of fluidizing gas expected with particles of Geldart's group B. It will be noted that as the bed passes from the quiescent state to the bubbling regime at $U/U_{mf} = 1$, there is a sharp rise in heat transfer coefficient as the particles become mobile. The bubbling action induces mixing of particles throughout the bed, transporting particles to and from the surface and the bulk of the bed. However, in the bulk of the bed, the particles exchange heat with the fluidizing gas and each other by conduction through the gas and usually stay there for a time sufficient to reach the bulk bed temperature.

As the velocity of the fluidizing gas is increased progressively, the heat transfer coefficient rises and reaches a maximum when U/U_{mf} has a value in the region of 1.5 to 2. The heat transfer coefficient thereafter gradually falls because the increase in bubble flow tends to cause sufficient 'blanketing' of the heat transfer surface by the fluidizing gas to counteract the effect of the reduced particle residence times produced by more violent bubbling.

Note that in the above discussion, the interphase gas convective component of heat transfer makes a contribution to the bed-to-surface heat transfer coefficient; in general, the gas convective component

80 Fluidized Bed Heat Transfer

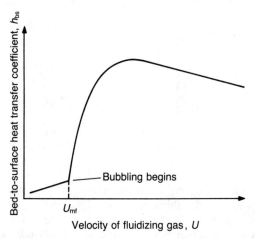

Figure 3.7 The bed-to-surface heat transfer coefficient as a function of fluidizing velocity—group B particles.

cannot be measured separately from the particle convective component, but is dominated by it for Geldart's category B particles. The approximate correlation of Zabrodsky *et al* (1976) for the value of the *maximum* value of the bed-to-surface heat transfer coefficient h_{max} with group B-type particles, which is given by equation (3.19), has been found to be one of the most reasonable predictions when the *bed temperature is less than 600 °C, the particle Reynolds number at minimum fluidizing velocity, Re_{mf}, is less than about 12.5 and Archimedes number Ar is less than about 26 000*. The temperature of 600 °C is that above which the radiative component of heat transfer starts to become significant, while the values of the dimensionless parameters are those corresponding to a change in fluidization characteristics of the bed mentioned in §2.4.1, following experimental findings by Botterill *et al* (1982). Under reasonable operating conditions, and subject to the above restrictions, Botterill (1983) and Botterill *et al* (1984) recommend that a value of about 70% of that predicted by equation (3.19) represents a conservative estimate for h_{max}. (Note that this equation is *not* dimensionless!)

$$h_{max} = 35.8 \rho_p^{0.2} k_f^{0.6} d_p^{-0.36}. \tag{3.19}$$

The units for equation (3.19) are SI units, namely $W\,m^{-2}\,K^{-1}$ for the heat transfer coefficient, $kg\,m^{-3}$ for density, $W\,m\,K^{-1}$ for thermal conductivity and m for particle diameter.

For beds behaving in the manner of Geldart's group D particles, for which Re_{mf} exceeds about 12.5 and Ar exceeds about 26 000, Botterill *et al* (1984) recommend for the maximum value of the particle convective heat transfer coefficient $h_{pc_{max}}$:

$$h_{pc_{max}} = 0.843 Ar^{0.15} k_f/d_p. \tag{3.20}$$

This is restricted to particle diameters exceeding 0.8 mm and bed temperatures less than about 600 °C; $Re_{mf} > \sim 12.5$, $Ar > \sim 26\,000$. For finer less-dense powders, corresponding to Geldart's group A, there is relatively little reliable information, particularly over a range of temperatures. The matter is complicated by expectations of changes in bed fluidization behaviour with temperature and by changes in the size distribution of group A powders due to attrition or elutriation, all of which can exert considerable influence on their bed-to-surface heat transfer performance.

It is in general wise to bear in mind the work of de Groot (1967), referred to in Chapter 2, which showed how different bubbling patterns can arise as a result of changes in particle size range as well as scale of equipment. This has implications for heat transfer because of its strong dependence upon bed behaviour.

3.2.4 Interphase gas convective component, h_{gc}

This component becomes important when the mean size of the bed particles is large, 0.8 mm upwards, and also when the fluidizing gas is at an elevated static pressure, which leads to the gas flow regime in the spaces surrounding the particles being transitional or turbulent, rather than laminar.

The empirical correlation due to Denloye and Botterill (1978) would seem to be the most useful in the present state of knowledge:

$$h_{gc} d_p^{0.5}/k_f = 0.86 Ar^{0.39}. \tag{3.21}$$

Notice that equation (3.21) is not dimensionless; the dimensions of the left-hand side (and, by equality, the right-hand side) clearly being (length)$^{-0.5}$, yielding units m$^{-0.5}$.

Equation (3.21) should be used cautiously of course, for it is based upon the results of experiments having a fairly close size range. It has been suggested by Botterill (1983) that the presence of fines among the particles may tend to inhibit turbulence within the gas stream, which helps explain results reported by Golan *et al* (1979) in which, with various large mean sizes of particles, changes of particle size exerted little effect on heat transfer.

3.2.5 Radiative component, h_{rad}

An estimation of the radiative heat transfer coefficient h_{rad} may be made using a combination of equations (3.2) and (3.3), but with the pertinent temperatures being those of the bed, T_b, and surface, T_s, thus

$$h_{rad} = \frac{\varepsilon_m \sigma (T_b^4 - T_s^4)}{(T_b - T_s)} \qquad (3.22)$$

where ε_m is an apparent value of emissivity of the surface, taken as approximately 0.6, and σ is the Stefan–Boltzmann constant.

Use of this apparent emissivity represents a 'rule of thumb' which has arisen from the work of Baskakov *et al* (1973), the problem being that the temperature of particles immediately adjacent to the heat transfer surface will be much lower than the bulk bed temperature because they have already exchanged heat (see Makhorin *et al* 1978, Borodulya *et al* 1984).

However, considerably more investigative work remains to be done before accurate assessments of the radiative component of heat transfer and its interaction with the particle convective component can be made. Botterill (1975, 1983) has pointed out that there is evidence that these two components are not completely independent of each other such that the radiative mode of heat transfers some heat at the expense of the particle convective mode and that this effect is greater with larger particles than with smaller ones.

It may be helpful to demonstrate the complexity of the situation. Consider the sequence of illustrations in figure 3.8. Figure 3.8(*a*) shows a 'packet' of hot particles which has just arrived at a cool plane surface or wall. If the surface of the particles is of zero emissivity, no radiative component of heat transfer arises and the entire heat transfer is by conduction through the gas between sector AB of the surface of any particle in contact with the wall (see figure 3.8(*b*)). If, then, the emissivity of the surface of the particles is increased, then those particles which are 'visible to the wall' start to lose heat to the wall by radiation. Accordingly, the temperature of these particular particles will start to fall, while the temperature of the particles in the bulk of the bed will remain unchanged. Further, if we consider a single particle in contact with the wall again, figure 3.8(*c*), the entire particle surface will now be radiating heat as well as losing heat, conducted away through the gas film adjacent to sector AB. Consequently, the surface temperature of the particle is lower and distributed differently from that in figure 3.8(*b*), so that, although the total heat transfer rate from the particle has increased, a smaller fraction of it is due to conduction through the gas. Finally, since the entire surface area of the particle radiates, the larger the particle is, the smaller the fraction of its surface area (sector AB) from which heat is conducted through the gas becomes, as indicated in figure 3.8(*d*). Thus, the larger the particle, the more the radiative mode transfers heat at the expense of conduction through the interstitial gas.

Heat Transfer in Beds of Particles 83

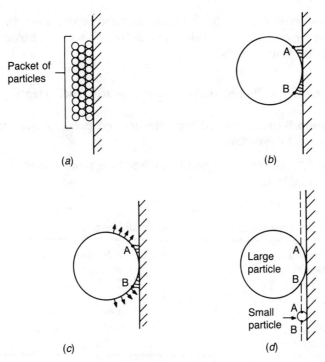

Figure 3.8 Illustrating the influence of radiative heat transfer between a particle and a surface and the effect of particle size thereon.

There is additional complexity when a bubble is adjacent to the wall, as shown in figure 3.9. The wall then has sight of a greater surface area of particles than when they are arranged in contact with the wall as shown in figure 3.8(a), but the bubble blankets off the wall from particle convection.

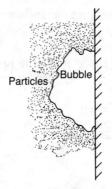

Figure 3.9 The influence of a bubble on the particle surface area visible to a wall.

All these effects are difficult to quantify and separate out by working from first principles and much research remains to be carried out to elucidate the matter further.

3.3 Estimation of Bed-to-surface Heat Transfer Coefficients

3.3.1 Group B particles; bed temperature and pressure near to ambient conditions

Example 3.2. Estimate the bed-to-surface heat transfer coefficient given the data in table 3.2.

Table 3.2 Data for group B particles and fluidizing gas.

Mean particle size, d_p	0.485 mm
Particle density, ρ_p	2600 kg m^{-3}
Bed temperature, T_b	227 °C
Fluidizing gas properties at bed temperature:	
Thermal conductivity, k_f	4.041×10^{-5} kW m^{-1} K^{-1}
Viscosity, μ_f	2.67×10^{-5} kg m^{-1} s^{-1}
Density, ρ_f	0.706 kg m^{-3}

First, check whether Reynolds number Re_{mf} at minimum fluidizing velocity, equation (2.41), is less than about 12.5 and Archimedes number Ar, equation (2.44), is less than about 26 000, as discussed in §3.2.3.

This raises the problem of determining the minimum fluidizing velocity U_{mf}. It is preferable to determine U_{mf} experimentally, as outlined in §2.4.4. Suppose that this has been done and U_{mf} has been found to be 0.15 m s^{-1}. (The alternative is to use an empirical correlation, e.g. equation (2.48), which may give a sufficiently close approximation, provided that the resulting value of Re_{mf} is considerably smaller than about 12.5.)

$$Re_{mf} = \frac{0.706 \times 0.15 \times 0.485 \times 10^{-3}}{2.67 \times 10^{-5}} \tag{3.23}$$

$$= 1.924$$

$$Ar = \frac{0.706 \times (2600 - 0.706) \times 9.81 \times (0.485 \times 10^{-3})^3}{(2.67 \times 10^{-5})^2} \tag{3.24}$$

$$= 2879.$$

Now use Zabrodsky's equation (3.19) to obtain the maximum value of the bed-to-surface heat transfer coefficient h_{max}

$$h_{max} = 35.6 \times (2600)^{0.2} \times (4.041 \times 10^{-2})^{0.6} \times (0.485 \times 10^{-3})^{-0.36}$$
$$= 390 \text{ W m}^{-2}\text{ K}^{-1}.$$

As recommended in §3.2.3, a value of about 70% of this may be expected, so that the bed-to-surface heat transfer coefficient h_{bs} will be given by

$$h_{bs} = 0.7 \times 390$$
$$= 273 \text{ W m}^{-2}\text{ K}^{-1}.$$

Note, here, that the particle convective component and interphase gas convective components of heat transfer have not been separated out, so that h_{bs} includes the effect of both components. It is necessary to separate out these components with group D particles or when the gas pressure is elevated, as examples in the next section show.

3.3.2 Group D particles; bed temperature and pressure near to ambient conditions

Example 3.3. Estimate the bed-to-surface heat transfer coefficient given the data in table 3.3

Table 3.3 Data for group D particles and fluidizing gas.

Mean particle size, d_p	2.0 mm
Particle density, ρ_p	2600 kg m^{-3}
Bed temperature, T_b	227 °C
Fluidizing gas properties at bed temperature:	
Thermal conductivity, k_f	4.041×10^{-5} kW m^{-1} K^{-1}
Viscosity, μ_f	2.67×10^{-5} kg m^{-1} s^{-1}
Density, ρ_f	0.706 kg m^{-3}

First, check Re_{mf} and Ar, equations (2.41) and (2.44). The minimum fluidizing velocity U_{mf}, however, needs to be known, preferably by experimental determination. Suppose this has been determined at bed temperature and found to be 0.97 m s^{-1}

$$Re_{mf} = \frac{0.706 \times 0.97 \times 2.0 \times 10^{-3}}{2.67 \times 10^{-5}} \tag{3.25}$$
$$= 51.3$$

$$Ar = \frac{0.706 \times (2600 - 0.706) \times 9.81 \times (2.0 \times 10^{-3})^3}{(2.67 \times 10^{-5})^2} \tag{3.26}$$
$$= 202\,000.$$

86 Fluidized Bed Heat Transfer

These values of Re_{mf} and Ar considerably exceed 12.5 and 26 000, respectively. In consequence, the particle convective and interphase gas convective components of heat transfer must be calculated separately using the correlations given by equations (3.20) and (3.21)

$$h_{pc_{max}} = \frac{0.843 \times (202\,000)^{0.15} \times (4.041 \times 10^{-2})}{2.0 \times 10^{-3}} \quad (3.27)$$

$$= 106 \text{ W m}^{-2}\text{K}^{-1}$$

$$h_{gc} = \frac{0.86 \times (202\,000)^{0.39} \times (4.041 \times 10^{-2})}{(2.0 \times 10^{-3})^{0.5}} \quad (3.28)$$

$$= 91.1 \text{ W m}^{-2}\text{K}^{-1}.$$

Finally, from equation (3.18), neglecting the radiative term because the bed temperature is less than 600 °C, the bed-to-surface heat transfer coefficient h_{bs} is

$$h_{bs} = 106 + 91.1$$
$$= 197.1 \text{ W m}^{-2}\text{K}^{-1}.$$

3.3.3 Group B and D particles; effect of high bed temperature and pressure

The viscosity and thermal conductivity of the fluidizing gas change significantly with temperature; generally, both increase (see any table of gas properties). The gas density increases with pressure, but reduces with temperature (see, for example, the characteristic equation for a perfect gas, $pV = mRT$).

These gas properties are parameters in the Reynolds and Archimedes numbers, equations (2.41) and (2.44). It will be demonstrated here by numerical example that, with group D-type particles, the Reynolds and Archimedes numbers can decrease with temperature until they become less than the values of 12.5 and 26 000 at which the flow through the voids changes from turbulent/transitional, when the interphase gas convective component of heat transfer makes a significant contribution, to laminar, when it becomes very small and the thermal conductivity of the gas is the overriding gas property.

Example 3.4. Estimate the bed-to-surface heat transfer coefficient at the bed temperatures given with the data in table 3.4.

Consider the first column of data in table 3.4 and determine the Archimedes Number Ar

$$Ar = \frac{0.706 \times (2600 - 0.706) \times 9.81 \times (1.3 \times 10^{-3})^3}{(2.67 \times 10^{-5})^2} \quad (3.29)$$

$$= 55\,479.$$

Estimation of Bed-to-surface Heat Transfer Coefficients

Table 3.4 Data for a bed of group D particles at different temperatures.

Mean particle size, d_p		1.3 mm
Particle density, ρ_p		2600 kg m^{-3}
Shape factor, φ, determined by experiment		0.7
Bed temperature, T_b (°C)	227	877
Fluidizing gas properties:		
Thermal conductivity, k_f (kW m^{-1} K^{-1})	4.041×10^{-5}	7.427×10^{-5}
Viscosity, μ_f (kg m^{-1} s^{-1})	2.67×10^{-5}	4.511×10^{-5}
Density, ρ_p (kg m^{-3})	0.706	0.3069
Observed bed voidage at minimum fluidization, ε_{mf}	0.45	0.45

The minimum fluidizing velocity is not given, so that Re_{mf} cannot be determined directly using equation (2.41). However, if the particle shape factor φ and the bed voidage at minimum fluidizing velocity, ε_{mf}, are known accurately, it is possible to calculate Re_{mf} using equation (2.45), which was derived from the Ergun equation (2.40). Purely to illustrate that an increase of bed temperature could cause the gas flow through the voids to change regime, it is assumed here that ε_{mf} is known accurately, although in practice it may not have been determined. Equation (2.45) is rearranged below as equation (3.30):

$$aRe_{mf}^2 + bRe_{mf} - Ar = 0 \tag{3.30}$$

where

$$a = 1.75/\varphi\varepsilon_{mf}^3 \tag{3.31}$$

$$b = 150(1 - \varepsilon_{mf})/\varphi^2\varepsilon_{mf}^3. \tag{3.32}$$

Continuing, we have

$$a = 1.75/(0.7 \times 0.45^3) = 27.43$$

$$b = 150(1 - 0.45)/(0.7^2 \times 0.45^3)$$

$$= 1848$$

so that the quadratic equation (3.30) becomes

$$27.43\,Re_{mf}^2 + 1848\,Re - 55\,479 = 0. \tag{3.33}$$

The positive root is

$$Re_{mf} = \frac{-1848 + [1848^2 - 4 \times 27.43 \times (-55\,479)]^{1/2}}{2 \times 27.43}$$

$$= 22.5.$$

The flow through the voids is thus transitional/turbulent and group

D-type bed behaviour is to be expected.

Considering now the situation at 877 °C

$$Ar = \frac{0.3069 \times (2600 - 0.3069) \times 9.81 \times (1.3 \times 10^{-3})^3}{(4.511 \times 10^{-5})^2} \quad (3.34)$$

$$= 8450.$$

If the voidage at minimum fluidizing velocity, ε_{mf}, is unchanged, then the quadratic equation (3.33) in Re_{mf} now becomes modified to

$$27.43\ Re_{mf}^2 + 1848\ Re_{mf} - 8450 = 0 \quad (3.35)$$

and the corresponding positive root is

$$Re_{mf} = \frac{-1848 + [1848^2 - 4 \times 27.43 \times (-8450)]^{1/2}}{2 \times 27.43}$$

$$= 4.30. \quad (3.36)$$

Thus, the Reynolds number at minimum fluidizing velocity, Re_{mf}, has fallen below 12.5 because of the increase in bed temperature. Even if we allow for the fact that the voidage at minimum fluidization, ε_{mf}, is changed by say about 10% from the assumed 0.45 to either 0.49 or 0.41, thus altering the values of the coefficients a and b in equation (3.30), the reader is invited to verify that the same conclusion would still hold in this particular case.

Accordingly, the bed-to-surface-heat transfer coefficient h_{bs} at 227 °C has to be calculated using equations (3.20), (3.21) and (3.18), while that at 877 °C has to be calculated using equations (3.19), (3.22) and (3.18). Thus, at 227 °C

$$h_{pc_{max}} = \frac{0.843 \times (55\,479)^{0.15} \times (4.041 \times 10^{-2})}{(1.3 \times 10^{-3})}$$

$$= 135\ W\,m^{-2}\,K^{-1}$$

$$h_{gc} = \frac{0.86 \times (55\,479)^{0.39} \times (4.041 \times 10^{-2})}{(1.3 \times 10^{-3})^{0.5}}$$

$$= 68.3\ W\,m^{-2}\,K^{-1}$$

$$h_{bs} = h_{pc_{max}} + h_{gc}$$

$$= 135 + 68.3$$

$$= 203\ W\,m^{-2}\,K^{-1}.$$

At 877 °C, using equation (3.19) for the non-radiative component

$$h_{max} = 35.8 \times (2600)^{0.2} \times (7.427 \times 10^{-2})^{0.6} \times (1.3 \times 10^{-3})^{-0.36}$$

$$= 397\ W\,m^{-2}\,K^{-1}.$$

Taking the recommended 70% of this value gives a non-radiative heat transfer coefficient h_{nr}:

$$h_{nr} = 0.7 \times 397$$
$$= 278 \text{ W m}^{-2}\text{K}^{-1}.$$

An estimate of the radiative component of heat transfer using equation (3.22) requires that the value of emissivity ε_m and the temperature of the bed-to-surface interface be known. The recommended value for the former is about 0.6. The temperature of the bed-to-surface interface, T_s, depends upon the ratio of the thermal resistance on the bed side of the surface to the overall thermal resistance between bed and coolant. Suppose, for brevity here, that the temperature $T_s = 150\,°\text{C}$ (423 K), and take the Stefan–Boltzmann constant σ as 5.67×10^{-8} W m^{-2}K^{-4}. Inserting these values into equation (3.22) gives

$$h_{rad} = \frac{0.6 \times (5.67 \times 10^{-8}) \times (1150^4 - 423^4)}{(1150 - 423)}$$
$$= 80.3 \text{ W m}^{-2}\text{K}^{-1}.$$

Adding these components together gives

$$h_{bs} = 278 + 80.3$$
$$= 358 \text{ W m}^{-2}\text{K}^{-1}.$$

If the pressure is then elevated, the fluidizing gas density increases and, with it, the Archimedes number and the Reynolds number at minimum fluidization, as calculated above. If in the above example the pressure was raised sufficiently at 877 °C, then it can be envisaged that the Reynolds number could be raised above 12.5 with attendant effects upon bed behaviour and the bed-to-surface heat transfer coefficient.

3.4 Heat Transfer between the Bed, Distributor, Containing Walls, Immersed Tubes or Components

3.4.1 Bounding surfaces
The bed is bounded by the distributor at the bottom, and by the side walls while at the top; heat can be transferred to or from the free surface. Heat will be transferred wherever a temperature difference exists between the bed and bounding surfaces; heat transfer across the free surface will, however, be discussed in §3.5.

3.4.2 Distributor
For purposes of discussion figure 3.10 shows three alternative types of distributor, namely a porous plate type (figure 3.10(a)), a 'stand-pipe' type (figure 3.10(b)) and a simple perforated plate type (figure 3.10(c)).

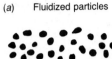
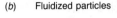

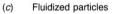
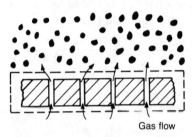

Figure 3.10 Alternative types of distributor—the choice affects the heat transfer between the bed and distributor. (*a*) Porous plate, (*b*) stand-pipe type, (*c*) perforated plate.

Heat can be transferred to or from the distributor material through any part of its surface, internal or external. Suppose, first, that the fluidizing gas is cooler than the bed particles. If a 'system boundary' is drawn around the outside of the plate, as shown in each of figures 3.10, it will be seen that heat will be transferred to the upper surface of the distributor from the bed, while the lower surface and internal surfaces will be cooled by the fluidizing gas. The rate of heat transfer to each surface depends upon the local mechanism of heat transfer and the amount of surface area exposed to it. Thus, the porous plate, figure 3.10(*a*), with its very large internal surface area, will be cooled by the gas much more than the simple perforated plate, figure 3.10(*c*), which has relatively little surface area. However, particles can accumulate on the plate between the holes and tend to insulate the plate from the bed. Wen *et al* (1980) have investigated some of the circumstances under which these 'dead zones' can be eliminated or promoted, and caution needs to be exercised in judging likely conditions at the upper surface of this type of distributor. The 'stand-pipe'-type distributor shown in figure 3.10(*b*) has an intermediate amount of internal surface; but even when the bed is fluidized, a quantity of particles can settle permanently on to the plate between the stand pipes and form a stable insulating layer between the mobile particles of the bed and the plate. This type of distributor, therefore, tends to receive less heat from the bed and its material operates at a less severe temperature. Proper thermal analysis is therefore required for the design of distributors, to establish likely working temperatures of the material, the tendency for thermal distortion etc, which may lead to mal-distribution of fluidizing gas, leakage of seals etc.

3.4.3 Side walls

Conditions at the containing walls of a fluidized bed are not the same as in the freely bubbling regions several particle diameters within the bed. The flow of particles at the wall surface tends to be generally downwards, the particles sticking and slipping down intermittently; the particle flow at the containing wall is only occasionally disturbed by bubbles which move to the wall. This general situation is illustrated by figure 3.11. Because of comparatively little particle convection and the relatively long particle residence time at any one region on the containing wall (as can be observed by a simple laboratory experiment with particles fluidized within a transparent tube), the heat transfer coefficient at the bed-to-containing-wall interface may be expected to be lower than that to, say, a horizontal cooling tube immersed in the well fluidized region of the bed. The model due to Yoshida *et al* (1969) implies that a downward flow of solids at the wall constitutes a thermal boundary layer similar to that when an inviscid fluid slides down the wall. There are practical difficulties associated with using such models, as pointed out by Glicksman (1984), in that the mathematical expression

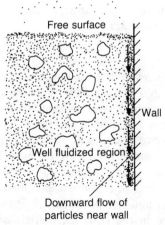

Figure 3.11 Illustrating particle flow at a containing wall.

for the heat transfer coefficient arising from such modelling requires knowledge of an 'effective' thermal conductivity, k_e, of the emulsion sliding down the wall, and the time taken to slide down, t_s. The latter can only be estimated crudely. Furthermore, knowledge of the fraction of the wall covered by bubbles and the velocity of sliding is also required. The best that can be done in the absence of such information seems to be that suggested by Glicksman (1984), namely to use equation (3.37) to provide a minimum value for the bed-to-wall heat transfer coefficient h_{bw}, thus

$$h_{bw} = \left(\frac{4k_e \rho_g c_g U_{mf}}{\pi H}\right)^{1/2} \quad (3.37)$$

where ρ_g is the gas density, c_g is the gas specific heat and H is the bed height.

Thus, adequate models for the wall heat transfer situation have not yet been fully developed, while the correlations given in the previous sections of this chapter were obtained from surfaces immersed in the freely bubbling region of the bed and cannot, therefore, be applied to estimate heat transfer at the vertical containing wall with any great confidence. Experiments with the particular particles and bed involved in obtaining the required data are therefore to be recommended.

Relatively little experimental data on heat transfer across the vertical containing walls of fluidized beds have been reported in the literature and even this has been in small-scale apparatus. Hoebink and Rietema (1978) report experiments with spent cracking catalyst particles of mean size about 66 μm (note that these are Geldart's group A particles, about which heat transfer data are relatively scant). These experiments show that (i) the heat transfer coefficient at the bed–wall interface depends upon the depth of the bed because of the downward movement of the particles and (ii) that the local heat transfer coefficient tends to decrease with height above the distributor.

3.4.4 Immersed tubes

Frequently, additional surface has to be provided in order to transfer heat to or from a fluidized bed, perhaps because of insufficient area of containing wall or for constructional reasons. Commonly, tubes carrying the heat transfer fluid are installed in the bed. The presence of tubes in the bed, used to remove heat, will influence the local fluidization pattern as will other obstructions such as baffles, used to break up bubbles. This has particular importance for particle convection, which plays such an important role in bed-to-surface heat transfer. Clearly, the size and shape of tubes and the presence of fins or extended surfaces, the location of the tube in the bed, their orientation, particle size and size range, gas properties etc all influence the magnitude of the

Heat Transfer Between Bed, Distributor, Walls, Tubes or Components 93

bed-to-surface heat transfer coefficient. It is only possible here to make a few comments about some of these features.

Flow pattern around a single horizontal tube
Figure 3.12 shows a sketch of the particle, bubble, and emulsion pattern in the locality of a single horizontal tube. The important features are:

(a) the tendency for a stagnant heap of de-fluidized particles to settle over part of the top surface of the tube, inhibiting local heat transfer;

(b) the lower surface of the tube is subjected to alternate contact with bubbles and particles transported by the bubbles—see the illustration of a rising bubble about to contact the lower surface;

(c) the side surfaces tend to become washed by particles streaming past.

Accordingly, the local heat transfer coefficient is non-uniformly distributed around the tube surface (see Botterill *et al* 1984) but heat transfer estimated from correlations averages out these effects.

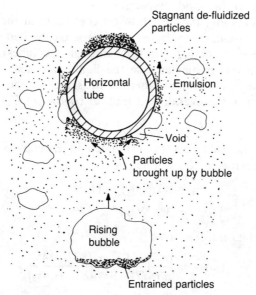

Figure 3.12 Conditions near a tube immersed in a fluidized bed.

Estimation of heat transfer coefficients
The correlations used as examples in previous sections of this chapter are applicable to a single horizontal tube. Banks of horizontally placed tubes, whether arranged in line or in staggered rows, show similar heat

transfer coefficients which reduce somewhat as the distance between adjacent tube surfaces is reduced (see, for example, Maclaren and Williams 1969). When the tubes are close together, however, there can be a sharp reduction in heat transfer coefficient due to the restriction of particle convection.

Example 3.5. A fluidized bed has to be maintained at a temperature of 877 °C, the relevant data being given in table 3.4. Estimate the length of immersed tube of 25 mm bore required to extract heat from the bed at the rate of 40 kW, given that the cooling medium passing through the tube is liquid which enters at 45 °C and leaves at 95 °C.

Appropriate correlations for the determination of the heat transfer coefficient at the liquid-to-tube-surface interface may be found in texts on convective heat transfer (see, for example, McAdams 1954). Such correlations are of the form given in equation (3.38) and heed has to be taken of their range of applicability.

$$Nu = \text{constant} \times Re^n \times Pr^m \tag{3.38}$$

where Re and Pr are the Reynolds number for the fluid flow and the Prandtl number of the fluid, respectively.

However, generally, the heat transfer coefficient at the liquid side of the tube, h_L, is significantly larger than those on the bed side (note that this is *not* the case for gases flowing through the tubes). Suppose, for brevity, that in this particular case the liquid side heat transfer coefficient is $6 \, \text{kW m}^{-2} \, \text{K}^{-1}$ and the thermal resistance of the tube wall may be neglected, then the overall heat transfer coefficient U_o is given by

$$U_o = [(1/h_L) + (1/h_{bs})]^{-1}. \tag{3.39}$$

Hence, using the value for h_{bs} estimated in Example 3.4, the overall heat transfer coefficient U_o will be

$$U_o = [(1/6000) + (1/358)]^{-1}$$

$$= 338 \, \text{W m}^{-2} \, \text{K}^{-1}.$$

The LMTD, equation (3.8), gives

$$\text{LMTD} = \frac{(877 - 95) - (877 - 45)}{\ln[(877 - 95)/(877 - 45)]}$$

$$= 806.7 \, \text{K}.$$

The heat transfer rate to the tube, $\dot{Q}$, is given by the well known expression for heat exchanger design:

$$\dot{Q} = U_o A \text{LMTD} \tag{3.40}$$

where A is the surface area of the tube and U_o is the overall heat transfer coefficient. Hence, the required surface area A is

$$A = \frac{40}{0.338 \times 806.7}$$
$$= 0.147 \text{ m}^2$$

so that the tube length L is

$$L = \frac{0.147}{\pi \times 0.025}$$
$$= 1.87 \text{ m}.$$

In the event of there being insufficient space to incorporate a tube of such length in the bed (even if the tube is formed into a hairpin or spiral) the required length can be reduced by using extended surface, i.e. finned, tubing, as discussed in §3.4.5.

3.4.5 Finned tubes

Heat transfer to the liquid cooled tube described in Example 3.5 was controlled primarily by the heat transfer coefficient at the outer surface because the bed-to-tube-surface heat transfer coefficient is much smaller than that on the liquid side of the tube. If extra area is provided on the controlling side of the tube, then the tube length required for a given duty would be significantly shorter. This may be illustrated by the first case investigated in Example 3.6. However, the presence of fins when spaced too closely together can inhibit particle convection with attendant reduction of the bed-to-surface heat transfer coefficient. The fin spacing at which this reduction becomes important is not completely resolved, but Al-Ali (1976) suggests that when the gap between fins becomes less than 20 particle diameters, the maximum heat transfer coefficient starts to decline rapidly, and when the spacing is less than 5 particle diameters, the particle motion is greatly restricted. This has consequences for the designer faced with the task of extracting as much heat as possible from a given volume of bed, so that the product of bed-to-surface heat transfer coefficient and the *effective* surface area packed into the bed can be maximized.

Example 3.6. The duty required of a finned tube of 25 mm bore and 35 mm outside diameter immersed in a fluidized bed is identical to that described in Example 3.5. The fins on the outside of the tube are of 60 mm diameter and 2 mm thickness and are arranged at a pitch of 17 mm, as shown in figure 3.13. Estimate the required length of finned tube, assuming that the thermal conductivity of the tube material is 45 W m^{-1} K^{-1} and

96 Fluidized Bed Heat Transfer

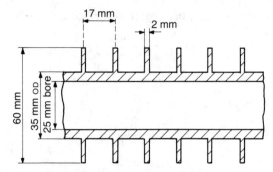

Figure 3.13 A diagram of finned tubing for Example 3.6.

(a) that the presence of the fins does not affect the bed-to-surface heat transfer coefficient given in Example 3.5, and alternatively,

(b) that the bed-to-surface heat transfer coefficient is only 50% of that given in Example 3.5.

First, estimate the effective area of the outer surface of the finned tube. Referring to figure 3.13, the area of fin in contact with the bed is termed the 'secondary area' A_s; if the tip area is neglected for one fin pitch length of tube of 17 mm

$$A_s = (\pi/4)(60^2 - 35^2) \times 2$$
$$= 3731 \text{ mm}^2.$$

The primary area between the faces of two fins, A_p, is

$$A_p = \pi \times 35 \times 15$$
$$= 1648 \text{ mm}^2.$$

Analyses of the thermal effectiveness of a variety of fin geometries were performed by Harper and Brown (1922) and Gardner (1945). They revealed that because of the finite thermal conductivity of the fin material, the surface is not isothermal. The usual way around this difficulty is to multiply the secondary area by a 'fin efficiency' η_f whose value is a function of the fin dimensions, the material thermal conductivity k, and the heat transfer coefficient, so that the effective heat transfer surface area A_e is given by

$$A_e = A_p + \eta_f A_s. \tag{3.41}$$

Charts and equations for the determination of the fin effectiveness η_f may be found in many of the standard books on heat transfer, e.g. Kreith (1965) and Schneider (1955), as well as the original papers by Harper and Brown and Gardner. For the purpose of this example the fin effectiveness η_f has been determined for the two different bed-to-surface heat transfer coefficients, namely 363 W m^{-2} K^{-1} and 50% of

Heat Transfer Between Bed, Distributor, Walls, Tubes or Components

363 W m^{-2}. The values are 0.65 and 0.75 respectively.

Case (a) h_{bs} = 363 W m^{-2} K^{-1}.

The effective area, A_e, on the bed side of the finned tube for *one fin pitch* of 17 mm is

$$A_e = 1648 + 0.65 \times 3731$$
$$= 4073 \text{ mm}^2.$$

The overall thermal resistance $(1/U_o A_e)$ is given by

$$\frac{1}{U_o A_e} = \frac{1}{h_{bs} A_e} + \frac{1}{h_i A_i} + \frac{t}{k A_m} \qquad (3.42)$$

where h_i is the heat transfer coefficient at the inner surface

A_i is the inner surface area

t is the tube wall thickness

k is the thermal conductivity

A_m is the mean area (half-way through tube wall)

U_o is the overall heat transfer coefficient based on the effective area of the outer surface.

Inserting values for one fin pitch length of tube leads to

$$A_i = \pi \times 25 \times 17$$
$$= 1335 \text{ mm}^2$$
$$A_m = \pi \times 0.5(25 + 35) \times 17$$
$$= 1602 \text{ mm}^2.$$

$$1/U_o A_e = 1/(0.363 \times 4073 \times 10^{-6}) + 1/(6 \times 1335 \times 10^{-6})$$
$$+ 0.005/(0.045 \times 1602 \times 10^{-6})$$
$$= 870.6 \text{ K kW}^{-1}$$

for one fin pitch. Hence the number of fins, N, required for a duty of 40 kW, using equation (3.40) with an LMTD of 806.7 K, is

$$N = \frac{40 \times 870.6}{806.7}$$
$$= 43.2$$

say 44 fins. The corresponding length of the finned tube, L, is then

$$L = 44 \times 0.017$$
$$= 0.748 \text{ m}$$

compared with 1.85 m if a plain tube is used.

Case (b) $h_{bs} = 181.5\ \mathrm{W\,m^{-2}\,K^{-1}}$.

Inserting values in equation (3.41) gives

$$A_e = 1648 + 0.75 \times 3731$$
$$= 4446\ \mathrm{mm}^2$$

for one fin pitch length of finned tube. Repeating the use of equation (3.42) gives

$$1/U_o A_e = 1/(0.1815 \times 4446 \times 10^{-6}) + 1/(6 \times 1335 \times 10^{-6})$$
$$+ 0.005/(0.045 \times 1602 \times 10^{-6})$$
$$= 1433\ \mathrm{K\,kW^{-1}}$$

for one fin pitch, giving the number of fins, N, required as

$$N = \frac{40 \times 1433}{806.7}$$
$$= 71$$

and the corresponding tube length L as

$$L = 71 \times 0.017$$
$$= 1.2\ \mathrm{m}.$$

Thus, if particle convection is partly suppressed by the spacing between the fins being too small, a greater length of finned tubing is required to achieve the same duty than if particle convection were not reduced. If the fin spacing were to be increased, a greater length of finned tubing would be required on account of the increase in primary area at the expense of secondary area. It is left as an exercise for the reader to establish the length of finned tubing required if the fin spacing were increased to 20 particle diameters, instead of the 11.3 particle diameters implied in this example (see the data in table 3.4).

3.4.6 Location of tubes in bed and effect of bed depth

If a tube is located at a level in the bed close to the distributor its heat transfer performance will differ from that obtained if it is placed at a higher level; generally, the heat transfer coefficients will be lower at the higher levels. Further, Elliott and Virr (1973), Al-Ali and Broughton (1977) and Virr (1983) have reported that the use of shallow fluidized beds with bare or finned tubes immersed in them can display higher bed-to-surface heat transfer coefficients than are obtained in deep beds. Figure 3.14 compares some heat transfer coefficients obtained with shallow fluidized beds under the supervision of the late Professor Douglas Elliott at the University of Aston in Birmingham with that

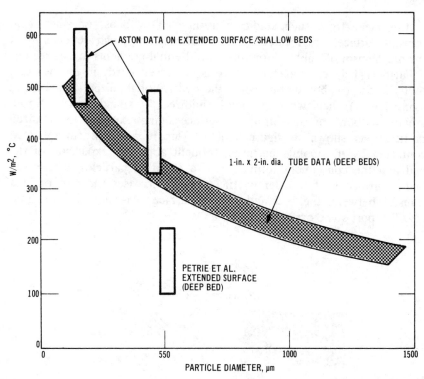

Figure 3.14 A comparison of the bed-to-surface heat transfer coefficient obtained in deep beds with those obtained in shallow beds. (After Elliott and Virr 1973.)

reported for deep beds. This reduction in performance when tubes are located in a deep bed is attributed mainly to bubble growth, so that a large fraction of the heat transfer surface can become engulfed in a bubble, as indicated in figure 3.15.

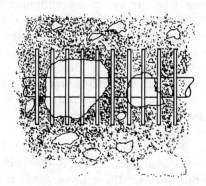

Figure 3.15 Large bubbles enveloping a finned tubing surface area.

100   Fluidized Bed Heat Transfer

It is clear from such studies that in shallow beds, or when heat transfer surfaces are located near the distributor, much depends upon the mechanism of bubble formation and thermal gradients in this region.

Further, if a heat transfer surface is immersed partly in the bed and partly in the freeboard zone when the bed is not fluidized, as shown in figure 3.16(a), then, when the bed is fluidized, it expands and leads to a greater amount of heat transfer surface being exposed to fluidized particles, as shown in figure 3.16(b). This increase in surface area contacted by the mobile particles, brought about by expansion of the bed and the consequent formation of a cloud of particles, affects the bed-to-surface heat transfer coefficient and increases the rate of heat transfer between the bed and such a surface. Al-Ali and Broughton (1977) report some experimental data.

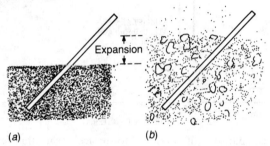

Figure 3.16 The effect of bed expansion on the surface area in contact with fluidized particles. (a) Static. (b) Fluidized.

3.5 Heat Transfer to Surfaces Located Above the Bed Free Surface

3.5.1 Classification of zones

The region above the free surface of an unfluidized bed of particles may, when the bed is fluidized, be considered as three distinct zones, although the precise boundary between each is not sharply defined. When fluidized, the bed expands, raising the level of the free surface, as shown in figure 3.16. The free surface level of the expanded bed may fluctuate between a minimum and maximum or, alternatively, remain fairly constant. The fluctuating situation tends to occur when the ratio of bed depth to diameter is about 1 or greater, e.g. when slugging (see Xavier and Davidson 1981).

Immediately above this expanded part of the bed there is the 'splash zone', while higher up still is the 'transport disengagement zone'. The main feature which distinguishes these three zones is the concentration of particles within them; thus, the largest number of particles per unit volume is found in the expanded part of the bed, a lesser concentration in the 'splash zone' and less still in the 'transport disengagement zone'. Although the heat transfer coefficients encountered in these zones are lower than in the bed, because of the reduced number of impacts with particles, each zone can contribute to heat exchange by particle convection to a surface located within it; the significant heat transfer zones, however, are the expanded and splash zones. This feature can be exploited by the designer to assist in the controlling of heat exchange between the bed and heat transfer surface over the operating range and, thus, has importance, for example, in the control of thermal output or 'turn down' of fluidized bed combustors. The reader is referred to Al-Ali and Broughton (1977), Xavier and Davidson (1981) and Chakraborty and Vickers (1984) for further information on such matters.

3.5.2 Radiation from the free surface

The free surface of a bubbling fluidized bed throws particles up when bubbles burst, so that a large surface area is available for radiation. Little in the way of quantitative study has been made of this phenomenon, but it is important in high-temperature beds since it can form a significant part of the heat loss from the bed, particularly if there is no reflection or back radiation from surfaces in the above-bed zones. Elliott and Virr (1973) briefly discuss radiation from the free surface of a shallow fluidized bed combustor fuelled with gas for potential use as a radiant heater, but elucidation of the subject remains to be done.

3.6 Concluding Remarks

It can be seen from the above that several mechanisms contribute to heat transfer in fluidized beds and that estimates of heat transfer coefficients depend heavily upon empirical correlations. Care has to be taken to ensure that such correlations are not applied outside their range of validity or conditions under which they were obtained. The scale of the equipment affects bed behaviour, which in turn affects heat transfer coefficients, thus be cautious in making predictions!

Examples

1. Hot air at atmospheric pressure (1.013 bar) and a temperature of 227 °C enters a bed of closely sized spherical particles of mean size 0.56 mm. The temperature of the particles is uniform throughout the bed at 77 °C. The bed is contained in a vessel of 1.3 m diameter, to a depth when fluidized of 30 mm. The mass flow of air through the bed is

1.11 kg s^{-1} and the properties of air at atmospheric pressure are given in the table below.

Temperature (°C)	77	127	177
Density (kg m^{-3})	1.008	1.014	1.021
Viscosity (kg m^{-1} s^{-1})	2.075 × 10^{-5}	2.268 × 10^{-5}	2.485 × 10^{-5}
Thermal conductivity (kW m^{-1} K^{-1})	3.003 × 10^{-5}	3.365 × 10^{-5}	3.710 × 10^{-5}

Determine the gas-to-particle heat transfer coefficient in the entry region of the bed using alternatively (*a*) equation (3.4) and (*b*) equation (3.17).

Make an estimate of the depth to which the gas must penetrate the bed before the gas temperature reaches 80 °C, using the two alternative values of heat transfer coefficient, if the voidage of the fluidized bed was 0.5. Comment on the different results predicted by these alternatives.

2. Estimate the bed-to-immersed-surface heat transfer coefficient for a bed of alumina particles operating at atmospheric pressure and a temperature of 477 °C, given the following properties:

    particle mean size   = 0.485 mm
    particle density   = 3750 kg m^{-3}
    fluidizing velocity   = 0.45 m s^{-1}.

Gas properties:
    density   = 0.4706 kg m^{-3}
    viscosity   = 3.482 × 10^{-5} kg m^{-1} s^{-1}
    thermal conductivity = 5.509 × 10^{-5} kW m^{-1} K^{-1}.

3. A fluidized bed reactor operates at a pressure of 8.5 bar and a temperature of 325 °C. Given the following data, estimate the bed-to-immersed-surface heat transfer coefficient:

    particle mean size   = 1.2 mm
    particle density   = 2640 kg m^{-3}.

Gas properties:
    density   = 5.0 kg m^{-3}
    viscosity   = 3.017 × 10^{-5} kg m^{-1} s^{-1}
    thermal conductivity = 4.661 × 10^{-5} kW m^{-1} K^{-1}.

Using the estimated bed-to-immersed-surface heat transfer coefficient, determine the surface heat flux if the surface temperature is maintained at a uniform temperature of 125 °C.

4. Heat is supplied to a fluidized bed by heated surfaces located within the bed. The particles are well mixed so that the bed temperature is uniform. Given the following data, estimate the heat transfer surface area required in the bed:

    bed temperature   = 102 °C.

Fluidizing gas properties at bed temperature:
    density   = 0.9413 kg m^{-3}
    viscosity   = 2.181 × 10^{-5} kg m^{-1} s^{-1}
    thermal conductivity = 3.186 × 10^{-5} kW m^{-1} K^{-1}.

Particle properties:
    mean size   = 0.58 mm
    density   = 2640 kg m^{-3}
    heat transfer
    surface temperature = 227 °C
    duty of heat
    transfer surface   = 50 kW.

5. A fluidized bed using particles of mean size 0.388 mm and of particle density 3700 kg m^{-3} operates at atmospheric pressure and at 927 °C. The bed is used to heat metal spheres which are suddenly immersed in it. Estimate the bed-to-immersed-surface heat transfer coefficient and the time taken to heat a sphere of 5.6 mm diameter from an initial temperature of 15 °C to 725 °C, assuming the thermal conductivity of the metal to be very large and the metal density to be 7800 kg m^{-3}.

6. Hot sand has to be cooled from 252 to 52 °C by fluidizing it with atmospheric air at 20 °C in an open channel of 300 mm width, in which 32 plain, water cooled tubes of 25 mm diameter are immersed. The fluidizing velocity (empty bed) is 0.3 m s^{-1}. The amount of sand to be cooled is 650 kg h^{-1}.

(*a*) Assuming a mean bed temperature of 152 °C and using the following data, estimate the bed-to-surface heat transfer coefficient and the overall heat transfer coefficient.

Air properties:
    mean specific heat capacity   = 1.09 kJ kg^{-1} K^{-1}
    density at 152 °C   = 0.8334 kg m^{-3}
    viscosity at 152 °C   = 2.3855 × 10^{-5} kg m^{-1} s^{-1}
    thermal conductivity at 152 °C = 3.5375 × 10^{-5} kW m^{-1} K^{-1}.

Particle properties:
    mean size   = 0.57 mm
    shape factor   = 0.73
    particle density   = 2640 kg m^{-3}.

(*b*) Assuming plug flow of solids in counterflow to the cooling water and neglecting heat removed by the fluidizing air, estimate the length of cooling tubes required, given the following further data:

water inlet temperature = 15 °C
water outlet temperature = 40 °C
specific heat capacity = 0.798 kJ kg^{-1} K^{-1}.

Heat transfer coefficient:
(water side of tubes) = 1 kW m^{-2} K^{-1}.

(c) Estimate the rate of heat transfer to the air and compare it with the rate of heat transfer to the sand.

7. Finned tubing has a primary surface area of 0.056 m^2 per metre length, a secondary surface area of 0.52 m^2 per metre length and a fin effectiveness of 0.75. Estimate the heat extraction rate a 380 mm length of this tubing would be capable of when immersed in a fluidized bed having particles of 0.42 mm mean size and operating at 127 °C, given the following data:

mean temperature difference
between fin surface and bed = 75 K.

Gas properties:
density = 0.7844 kg m^{-3}
viscosity = 2.485 × 10^{-5} kg m^{-1} s^{-1}
thermal conductivity = 3.710 × 10^{-5} kW m^{-1} K^{-1}.

In an endeavour to increase the heat transfer rate from the bed it was proposed to increase the surface area per unit length of tubing by reducing the spacing between the fins to 4 mm. Discuss the feasibility of this proposal.

References

Al-Ali B M 1976 *PhD Thesis* University of Aston in Birmingham, p169
Al-Ali B M and Broughton J 1977 Shallow fluidized bed heat transfer *Appl. Energy* **2** 101–14
Baskakov A P, Berg B V, Vitt O K, Fillippovsky N F, Kirakosyan V A, Goldobin J M and Maskaev V V 1973 *Powder Technol.* **8** 273
Borodulya V A, Kovensky V I and Makhorin K E 1984 in *Fluidization* ed. D Kunii and R Toei (New York: Engineering Foundation) p379
Botterill J S M 1975 *Fluid-bed Heat Transfer* (London: Academic) p167 *et seq*
—— 1983 in *Fluidized Beds: Combustion and Applications* ed. J R Howard (London: Applied Science) ch 1
Botterill J S M, Teoman Y and Yüregir K R 1982 The effect of operating temperature on the velocity of minimum fluidization bed voidage and general behaviour *Powder Technol.* **31** 101–10
—— 1984 in *Proc. XVIth Int. Centre for Heat and Mass Transfer Int. Symp., Dubrovnik, Yugoslavia, 1984* (Belgrade: ICHMT) paper 2.4

Chakraborty R K and Vickers M A 1984 in *Proc. 3rd Int. Conf. Fluidized Bed Combustion: is it Achieving its Promise? London, 1984* (London: Institute of Energy) paper DISC/33

Decker N A and Glicksman L R 1981 Conduction heat transfer at the surface of bodies immersed in gas fluidized beds of spherical particles *AIChE Symp. Ser. No* 208 **77** 339–49

Denloye A O O and Botterill J S M 1978 Bed to surface heat transfer in a fluidized bed of large particles *Powder Technol.* **19** 197–203

Elliott D E and Virr M J 1973 in *Proc. 3rd Int. Conf. Fluidized Bed Combustion, Environmental Protection Technol. Ser.* EPA-650/2-73-053 (North Carolina: EPA Technical Publications)

Gardner K A 1945 Efficiency of extended surfaces *Trans. ASME* **67** 621–31

Glicksman L R 1984 in *Fluidized Bed Boilers: Design and Application* ed. P Basu (Toronto: Pergamon) pp63–100

Golan L P, Cherrington D C, Diener R, Scarborough C E and Wiener S C 1979 *Chem. Eng. Prog.* **75** 63

Harper W P and Brown D R 1922 Mathematical equations for heat conduction in the fins of air-cooled engines *National Advisory Committee for Aeronautics Report* 158

Hoebink J H B J and Rietema K 1978 in *Fluidization* ed. J F Davidson and D L Keairns (Cambridge: CUP) pp327–32

Kreith F 1965 *Principles of Heat Transfer* (Scranton, Pennsylvania: International Textbook Company)

Kunii D and Levenspiel O 1969 *Fluidization Engineering* (New York: Wiley)

Maclaren J and Williams D F 1969 Combustion efficiency, sulphur retention and heat transfer in pilot-plant fluidized-bed combustors *J. Inst. Fuel* **42** 303–8

Makhorin K E, Pikashov V S and Kuchin G P 1978 in *Fluidization* ed. J F Davidson and D L Keairns (Cambridge: CUP) pp93–7

McAdams W H 1954 *Heat Transmission* (New York: McGraw-Hill)

Schneider P J 1955 *Conduction Heat Transfer* (Reading, MA: Addison-Wesley)

Virr M J 1983 in *Fluidized Beds: Combustion and Applications* ed. J R Howard (London: Applied Science) ch 11

Wen C Y, Krishnan R and Kalyanaraman R 1980 in *Fluidization* ed. J R Grace and J M Matsen (New York: Plenum) pp405–12

Xavier A M and Davidson J F 1981 Heat transfer to surfaces immersed in fluidized beds and the freeboard region *AIChE Symp. Ser. No* 208 **77** 368–73

Yoshida K, Kunii D and Levenspiel O 1969 Heat transfer mechanisms between wall surface and fluidized bed *Int. J. Heat Mass Transf.* **12** 529–36

Zabrodsky S S, Antonishin N V and Parnas A L 1976 On fluidized bed-to-surface heat transfer *Can. J. Chem. Eng.* **54** 52–8

Bibliography

Howard J R 1983 (ed.) *Fluidized Beds: Combustion and Applications* (London: Applied Science)

Kato K, Ito H and Omura S 1979 Gas-particle heat transfer in a packed fluidized bed *J. Chem. Eng. Japan* **12** 403–5

Petrie J C, Freeby W A and Buckingham J A 1968 In-bed heat exchangers *Chem. Eng. Prog.* **64** 45

Zabrodsky S S 1973 Compound heat exchange between a high temperature gas-fluidized bed and a solid surface *Int. J. Heat Mass Transf.* **16** 241–8

4 Design of Simple Fluidized Beds

4.1 Introduction

The fluidized beds considered in this book are those employed for processes requiring gas–solids contacting, in order to carry out chemical reactions or physical processes such as heating or cooling of particles, drying, mixing etc. The particles can thus be chemically active, inert or a mixture of the two types, as can be the fluidizing gas.

A complete system needed to carry out such operations comprises more than the fluidized bed reactor vessel, as indicated in figure 4.1. In general, the solids have to be fed into and out of the reactor, entrained particles have to be separated from the exhaust gas, sometimes such particles being recycled back to the reactor; heat transfer surfaces may have to be installed within the reactor, the solids and gases leaving the reactor may have to be cooled or heated; fans or compressors are required for pumping the gases, pumps for supplying coolant, instrumentation and controls are essential and the system may have to be integrated with other operations at the plant. These peripheral items, i.e. feeders, separators, fans or compressors, pumps, heaters, coolers and controls, can in practice be costly, pose as many problems and be just as essential to reliable operation of the plant as the reactor vessel itself, even influencing important details of the reactor design. Moreover, the energy input to the system, whether occasioned by burning of fuel, pumping of gases, transporting solids or driving ancillaries can contribute greatly to the running cost, and influence the economics of the plant significantly.

It should always be remembered that the specification and contract for a commercial process plant requires those responsible for design and construction to give guarantees concerning rated performance, plant life, reliability, safety, emission of pollutants etc, coupled with a requirement to meet economic criteria with minimum technical and commercial risks. This can present the designer or project manager with a dilemma because the available data or the techniques for making such estimates are not always adequate. The deficiency has to be made up by drawing upon practical experience, in order to help make the judgements.

108 Design of Simple Fluidized Beds

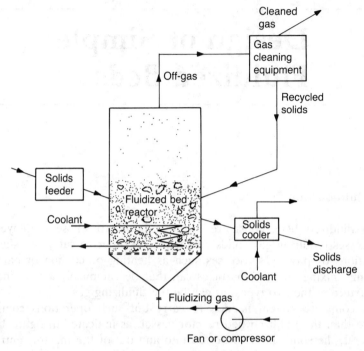

Figure 4.1 A fluidized bed plant.

In the case of fluidized bed systems, the prediction of behaviour of fluidized solids, the rates of heat and mass transfer, chemical reaction rates, mixing patterns and the distributions of reactant concentrations within a fluidized bed is difficult and often of insufficient precision or certainty. This situation arises partly because of an incomplete understanding of the phenomena and partly because of the limited scope of present-day mathematical models. Such models of the reactor or plant often have to be based upon a system of equations which are an insufficient description of behaviour; further, the values of the data to be inserted for solution of the equations may not be known, or be of insufficient accuracy. This is not to say that mathematical models should not be used for design purposes, but merely that their predictions should be viewed cautiously until verified by practical tests. Thus, the design of a fluidized bed reactor, particularly a large one, remains something of an art and has to rely considerably upon the data and experience obtained from pilot plant trials and elsewhere. (It should be noted that, in general, pilot plant needs to be as large as can be afforded, in order to reduce the uncertainties of extrapolation.) Data obtained from pilot plant trials may be inserted into an appropriate mathematical model which, if the agreement between that model and reality seems good, may be used for predictive purposes. Under such circumstances, however, it must be remembered that the model has only been calibrated against that particular pilot plant. The designer will still have to make a rough estimate of the dimensions of the reactor vessel in order that an initial estimate of costs may be made. In the next sections we consider elementary numerical examples which are intended to illustrate how the likely diameter, depth and fluidizing velocity may be estimated roughly, and then show the extent to which these estimated values will be altered by a change in constraints.

4.2 Estimation of Bed Dimensions and Fluidizing Velocity

Example 4.1. 10^4 kg h^{-1} of particulate solids, whose bulk density when loosely packed is 1600 kg m^{-3}, have to be processed (for example, dried, heated or reacted chemically) by fluidizing them using a gas whose mass flow rate is 2.16×10^4 kg h^{-1} and density at bed temperature is 0.7 kg m^{-3}. A fluidized bed is to be used for the purpose. Plant operational requirements demand that at least a one-hour supply of particles is contained in the bed.

Pilot plant experiments have shown that the minimum fluidizing velocity of the solids at bed temperature is 0.15 m s^{-1} and that excessive elutriation occurs if the velocity of the fluidizing gas exceeds 1.8 m s^{-1}.

Explore the possibility of meeting the following alternative criteria: (a) diameter = depth; (b) pumping power limited to 80 kW; (c) gas residence time to be not less than 0.8 s, i.e. in order to obtain a high degree of conversion of gas; and make a rough estimate of the corresponding bed diameters, depths and gas velocities, neglecting, for the sake of simplicity, the transport disengagement height required.

The system is shown schematically in figure 4.2. The operational requirement that the bed should contain a one-hour supply of solids leads to the volume V_p, required to contain the solids, being at least that computed by the following equation:

$$V_p = \frac{10^4 \times 1}{1600} \tag{4.1}$$

$$= 6.25 \text{ m}^3.$$

Criterion (a)

$$D = H. \tag{4.2}$$

In attempting to use equation (4.2), the first difficulty is that the bed depth H, when fluidized, depends upon the depth when not fluidized and the amount by which the bed expands as the velocity of the

110 Design of Simple Fluidized Beds

fluidizing gas is increased. Although relationships such as equation (4.3) below are often used to estimate bed expansion, they depend upon knowing the mean bubble size d_b, and this value is uncertain

$$\frac{H}{H_{mf}} = 1 + \frac{U - U_{mf}}{0.71\sqrt{gd_b}}. \tag{4.3}$$

It is also important to remember that the level of the free surface of the bed fluctuates continuously because of bubbling action, often by a considerable amount. If allowance has to be made for bed expansion, the safest procedure is to measure its extent and variation by pilot plant experiments and express the results as an empirical correlation, e.g. in the form

$$H = H_{mf}(1 + f(U)) \tag{4.4}$$

where $f(U)$ is some function of fluidizing velocity.

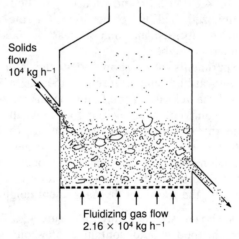

Figure 4.2 A schematic diagram of the fluidized bed used in Example 4.1.

Since, however, the purpose of this example is to illustrate the effect of changes in design criteria on size, it will be adequate in this instance only to put $H = H_{mf}$. Thus the diameter and depth of the bed become

$$D = H_{mf} = (4V_p/\pi)^{1/3} \tag{4.5}$$

$$= (4 \times 6.25/\pi)^{1/3}$$

$$= 1.99 \text{ m}$$

say 2 m. The fluidizing velocity U is given by

$$U = \frac{\dot{m}_f}{(\pi/4)D^2\rho_f} \tag{4.6}$$

$$= \frac{4 \times 2.16 \times 10^4}{\pi \times 3600 \times 0.7 \times 2^2}$$

$$= 2.73 \text{ m s}^{-1}. \tag{4.7}$$

It will be noted that this velocity exceeds that at which elutriation becomes excessive, (a value which also has to be established from experiment), so that the criterion $D = H$ cannot be met in this particular case. A larger diameter vessel is therefore required to accommodate the fluidizing gas flow. Since the plant specification fixes the bed volume, the bed will be correspondingly shallower.

Thus, using $U = 1.8 \text{ m s}^{-1}$ in equation (4.6)

$$D = \left(\frac{4 \times 2.16 \times 10^4}{\pi \times 0.7 \times 1.8 \times 3600}\right)^{1/2}$$

$$= 2.46 \text{ m}. \tag{4.8}$$

The bed depth

$$H_{mf} = \frac{6.25}{(\pi/4) \times 2.46^2} \tag{4.9}$$

$$= 1.31 \text{ m}.$$

Pumping power estimate at elutriation-limited fluidizing velocity.

For incompressible flow the power, P, required to pump the fluidizing gas through the system is the product of pressure drop Δp across the system and the volumetric flow rate:

$$P = \frac{(\Delta p)\dot{m}_f}{\rho_f}. \tag{4.10}$$

Neglecting pressure drops other than those across the bed and the distributor, the pressure drop Δp across the system becomes

$$\Delta p = \Delta p_b + \Delta p_d \tag{4.11}$$

where Δp_b and Δp_d are the pressure drops across the bed and distributor, respectively.

Since, when fluidized, Δp_b must be sufficient to support the weight of the bed

$$\Delta p_b = \frac{M_p g(\rho_p - \rho_f)}{(\pi/4)D^2\rho_p}. \tag{4.12}$$

Since the mass of particles in the bed, M_p, is given by

$$M_p = \rho_{bmf}\frac{\pi}{4}D^2 H_{mf} \tag{4.13}$$

112 Design of Simple Fluidized Beds

where ρ_{bmf} is the bed density at minimum fluidization and since the fluidizing gas density ρ_f is very small compared with that of the particles ρ_p, then from equations (4.12) and (4.13)

$$\Delta p_b = \rho_{bmf} g H_{mf}. \tag{4.14}$$

Hence

$$\Delta p_b = 1600 \times 9.81 \times 1.31$$
$$= 20\,560 \text{ N m}^{-2} \quad \text{or} \quad 20.56 \text{ kN m}^{-2}.$$

The pressure drop Δp_d across the distributor has to be sufficiently large to result in a uniformly distributed flow of gas into the bed. Different workers have suggested differing criteria for choosing Δp_d, see (a)–(c) below:

(a) $0.1 \Delta p_b$

(b) 0.35 m water gauge

(c) 100 times the loss of pressure due to the sudden expansion from inlet connection to the plenum.

It will be sufficient here, however, to assume that (b) gives a sufficiently large pressure drop, namely 0.35 m water gauge = 3.44 kN m^{-2}.

The pressure drop across the system, Δp, is then

$$\Delta p = 3.44 + 20.56$$
$$= 24.0 \text{ kN m}^{-2}.$$

From equation (4.10), the fluidizing gas pumping power is

$$P = \frac{24.0 \times 2.16 \times 10^4}{0.7 \times 3600}$$
$$= 206 \text{ kW}.$$

Criterion (b)

Pumping power, $P < 80$ kW.

From equation (4.10)

$$\Delta p < \frac{P \rho_f}{\dot{m}_f} \tag{4.15}$$

$$< \frac{80 \times 0.7 \times 3600}{2.16 \times 10^4}$$

$$< 9.33 \text{ kN m}^{-2}.$$

Hence, from equation (4.11), the allowable pressure drop across the

Estimation of Bed Dimensions and Fluidizing Velocity

bed, Δp_b, is

$$\Delta p_b = 9.33 - 3.44$$
$$= 5.89 \text{ kN m}^{-2}$$

and, from equation (4.14), the bed depth at minimum fluidizing velocity, H_{mf}, must not exceed

$$H_{mf} = \frac{5.89 \times 1000}{1600 \times 9.81}$$
$$= 0.375 \text{ m}$$

giving a bed diameter

$$D = \left(\frac{4 \times 6.25}{\pi \times 0.375}\right)^{1/2}$$
$$= 4.61 \text{ m.}$$

We need to check that the fluidizing velocity exceeds the minimum fluidizing velocity. (It is clear from comparison with equation (4.8) that this bed diameter will be sufficiently large to result in a fluidizing velocity of less than that giving excessive elutriation.)

From equation (4.6)

$$U = \frac{4 \times 2.16 \times 10^4}{3600 \times 0.7 \times \pi \times 4.61^2}$$
$$= 0.513 \text{ m s}^{-1}.$$

Discussion

The results of the above calculation are summarized in table 4.1 below. These demonstrate that achievement of the most compact reactor, $D = H$, may be thwarted by elutriation considerations and will be accompanied by relatively high pumping power demand. Conversely, designing to achieve a lower pumping power demand requires a shallower bed, leading to some sacrifice of compactness; in particular, a larger floor area is required to accommodate the larger diameter vessel.

Criterion (*c*)

Gas residence time to be not less than 0.8 s.

The fluidizing gas residence time t_{fr} is given by

$$t_{fr} = \frac{\varepsilon H}{U} \qquad (4.16)$$

where ε is voidage.

Once again we have the difficulty of not knowing by how much the

Table 4.1 The effect of changing design criteria on the size of a fluidized bed reactor.

Criterion	Diameter (m)	Height (m)	Power (kW)
Compactness, $D = H$ [a]	1.99	1.99	>206
Operation at elutriation limit	2.46	1.31	206
Reduced pumping power	4.61	0.375	80

[a] Not achievable due to the elutriation velocity being exceeded.

bed will expand with velocity, to permit an estimation of an appropriate value of H and H_{mf}, unless we have some data from pilot plant trials. (The results from pilot plant trials should still be used cautiously because, as the work of de Groot (1967) showed (see figure 2.10 in Chapter 2), the amount by which the bed expands depends upon the diameter of the bed and the particle size distribution.)

Nevertheless, if the bed is fluidized at the elutriation-limiting fluidizing velocity (1.8 m s^{-1}), and the voidage at this fluidizing velocity were, say, 0.6, then the minimum bed depth when fluidized is clearly

$$H = \frac{1.8 \times 0.8}{0.6} = 2.4 \text{ m} \tag{4.17}$$

whereas, if the velocity of the fluidizing gas was restricted to the minimum fluidizing velocity U_{mf}, the bed depth H_{mf} would then be (for $\varepsilon_{mf} = 0.42$)

$$H_{mf} = \frac{0.15 \times 0.8}{0.42} = 0.29 \text{ m}. \tag{4.18}$$

The large difference in depth between the values predicted by equations (4.17) and (4.18) makes it essential to obtain information about bed expansion if the minimum gas residence time criterion is to be implemented. To make progress, suppose that when fluidized at the elutriation-limiting velocity of 1.8 m s^{-1}, H/H_{mf} were to be 1.3, then from equation (4.17) the bed depth at the minimum fluidizing velocity H_{mf} would be $2.4/1.3 = 1.85$ m and the bed diameter and fluidizing gas pumping power would exceed that shown in the second row in table 4.1. It is left to the reader to explore the consequences of a greater or lesser amount of bed expansion on bed dimensions, fluidizing velocity and pumping power.

Summarizing all of the above shows the importance of deciding and specifying the appropriate criteria for design of the plant. Such a specification has to be based upon the particular plant economics, e.g. whether running costs dominate amortization of capital cost etc.

4.3 Transport Disengaging Height

For simplicity of illustration in Example 4.1 in §4.2, no consideration was given to the design of the zone above the free surface of the bed. When bubbles reach the free surface they burst, throwing particles upwards. Some of the fine particles become entrained in the fluidizing gas leaving the bed, while the coarser particles travel upwards at gradually reducing velocity and then fall back into the bed. A sufficient height of ducting above the bed must be provided to avoid transport of the coarser particles out of the system, since this constitutes a serious loss of product, and also to reduce the loss of fines. This was discussed briefly in §2.6.4 and shown in figure 2.17.

The transport disengaging height (TDH) is the term used to describe the necessary height of ducting required above the free surface of the bed to avoid excessive loss of solids. At the present time, it is not practicable to estimate the necessary TDH for given circumstances by working from first principles, nor are existing empirical correlations adequate for general application. There is simply insufficient understanding of the mechanisms and lack of data. Indeed, recently, Geldart (1985) pointed out that if the fluidized solids belong to group B or are a mixture having a wide size range, each case must be examined separately. Even with fine solids belonging to group A, where the greatest amount of research into elutriation has been done, the way ahead at the present time seems to lie in attempting to predict the entrainment rate versus freeboard height curve rather than calculating a TDH, which is ill defined.

The reader is referred to Geldart (1985) for a fuller treatment of elutriation and TDH; it is sufficient here to summarize by noting that there are two different definitions of TDH, namely

(i) the height, TDH(C), above the free surface of the bed required for the coarse particles flung up by bursting bubbles to disengage and fall back into the bed—above this height only fines are found; and

(ii) the height, TDH(F), at which the elutriation rate remains constant or declines only slightly.

These two heights are illustrated by figure 4.3 after Geldart (1985) who, although suggesting a procedure for the estimation of TDH(F) with group A particles, nonetheless urges the need to regard such predictions cautiously.

Once again, therefore, recourse must be made to previous experience of elutriation with plants processing the same particles with the same fluidizing gas, operating under similar conditions or carrying out tests with a pilot plant or a laboratory bed of sufficiently large diameter to be clear of the slugging regime.

116   Design of Simple Fluidized Beds

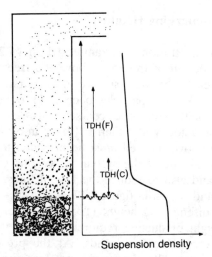

Figure 4.3 Transport disengagement heights. F = fines, C = coarse. (Redrawn from Geldart 1985.) Reproduced by permission of Academic Press.

Without such data, the specification of a TDH is entirely speculative. It is important to attempt to provide the right amount of TDH for providing acceptable solids loading and size distribution in the gas entering the downstream gas cleaning equipment. For example, insufficient TDH can result in an unnecessarily high solids concentration in the exhaust gas with attendant overloading of cyclones and escape of fines from the system; in addition, particle attrition in the exit ducts and unacceptable erosion of surfaces can arise. Too great a TDH increases the height and cost of the fluidized bed vessel. Although the height penalty can be reduced by increasing the diameter of the vessel in the above-bed zone, by virtue of the reduced gas velocity, this also increases cost, particularly if the operating pressure is high.

4.4 Distributors

Distributors were discussed qualitatively in §3.4.2, while figure 3.10 showed some alternative types. The intention here is to show how to estimate the sizes and distribution of holes for a distributor.

Example 4.2. A distributor is required for the fluidized bed process depicted in Example 4.1 when the bed is to operate at the elutriation limiting superficial fluidizing velocity of 1.8 m s^{-1}. The pressure drop at this velocity is not to exceed 0.35 m water gauge ($= 3.44 \text{ kN m}^{-2}$). Determine suitable hole diameters and number of holes for a simple perforated-plate-type distributor, given that the diameter and depth of the bed are shown in table 4.1 as 2.46 m and 1.31 m, respectively. The gas flow rate and density are $2.16 \times 10^4 \text{ kg h}^{-1}$ and 0.7 kg m^{-3}, respectively.

From fluid mechanics, the relationship between the volumetric flow rate $\dot{V}$ and the pressure drop Δp for incompressible flow through an orifice is given by

$$\dot{V} = \frac{C_d A_o (2\Delta p/\rho)^{1/2}}{[1 - (A_o/A_1)^2]^{1/2}} \qquad (4.19)$$

where A_o is the orifice flow area, A_1 is the flow area upstream of the orifice and C_d is the discharge coefficient of the orifice, i.e. the fluid density.

This equation may be rearranged in the form

$$\frac{A_o}{A_1} = \left(\frac{Q}{1+Q}\right)^{1/2} \qquad (4.20)$$

where

$$Q = \frac{\rho \dot{V}^2}{2 A_1^2 \Delta p C_d^2}. \qquad (4.21)$$

Inserting the given values in equation (4.21) gives

$$\dot{V} = \frac{2.16 \times 10^4}{3600 \times 0.7} \qquad (4.22)$$

$$= 8.571 \text{ m}^3 \text{ s}^{-1}$$

$$A_1 = (\pi/4) \times 2.46^2$$

$$= 4.753 \text{ m}^2$$

$$\Delta p = 3440 \text{ N m}^{-2}$$

while although the coefficient of discharge is a function of the Reynolds number of the flow, it will be sufficient for the present purpose to take the value $C_d = 0.6$. This gives

$$Q = \frac{0.7 \times 8.571^2}{2 \times 4.753^2 \times 3440 \times 0.6^2}$$

$$= 9.19 \times 10^4$$

and, hence
$$\frac{A_o}{A_1} = \left(\frac{9.19 \times 10^{-4}}{1 + 9.19 \times 10^{-4}}\right)^{1/2}$$
$$= 0.0303$$

i.e. 3.03% open area. If N is the number of holes in the distributor of diameter d_o and D is the bed diameter, then the two are connected by

$$Nd_o^2 = 0.0303 D^2. \tag{4.23}$$

Table 4.2 lists a few possible orifice sizes and numbers using equation (4.23).

Table 4.2 Diameters and corresponding numbers of holes required for the distribution of fluidized bed. (Pressure drop = 0.35 m water gauge, superficial fluidizing velocity = 1.8 m s^{-1}.)

Orifice diameter (mm)	Number required	Number cm^{-2}
0.5	7.3×10^5	15.4
1.0	1.83×10^5	3.85
1.5	8.13×10^4	1.71
2.0	4.75×10^4	0.963
3.0	2.03×10^4	0.427
4.0	1.14×10^4	0.240

The orifice diameter is constrained by two main factors; if the hole size is too small it can become blocked by dust, while if it is too large the smaller number of orifices and, hence, the larger distance between them can lead to a non-uniform distribution of gas and, under some circumstances, some flow-back of fine particles when the bed is slumped. In the latter event these particles accumulate in the plenum chamber (windbox) below the distributor and may cause erosion as they are blown about. Such considerations would probably lead to the chosen orifice diameter being 2 or 3 mm.

However, tests should be carried out with the particular particles concerned, to establish the extent of the flow-back of fines, before deciding the orifice size. If flow-back is excessive, a different design from the simple perforated plate, such as a bubble cap (see figure 4.4) may be required.

There are, however, several other considerations which affect the design of distributor and choice of type. For example, a large flat perforated plate will deflect under the weight of the particles contained in the bed and intermediate supports may be needed or the plate be made convex to provide greater rigidity than a flat plate. The distributor

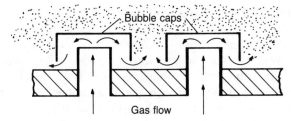

Figure 4.4 A bubble-cap-type distributor.

must be adequately sealed around its periphery so as to prevent leakage of gas and must be of a construction that will withstand the loads and operating conditions without distortion, thermal or otherwise. With high-temperature processes it may be necessary to protect the distributor from heating by the bed in which case a 'stand-pipe'-type distributor, such as that shown in figure 3.10(b), is one solution; indeed, it may be necessary to have a water cooled type of stand-pipe distributor, as shown in figure 4.5.

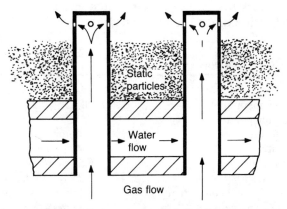

Figure 4.5 A water cooled distributor.

Sufficient thought is not always given to the problem of manufacture; for example, the use of a pre-perforated plate having hole spacings and sizes near to the desired values might be adapted satisfactorily, and may be cheaper and take less time to make than a purpose-made plate with individually drilled holes. Further, when it comes to large sizes, the distributor may have to be made in sections; then it must be considered how the plate or sections are to be lifted and installed without damage. Such matters may seem mundane and somewhat removed from fluidization phenomena, but they have to be faced and solved by the designer and contractor if the plant is to be built.

4.5 Heat Removal from Fluidized Beds

Some processes carried out in fluidized beds are exothermic and, in order to prevent excessive temperatures arising, cooling surfaces have to be located within the bed. In other processes, hot solids may have to be cooled and heat removal through the containment walls may require supplementing by immersing cooling tubes in the bed. This section demonstrates how cooling surface areas may be estimated.

Example 4.3. A fluidized bed reactor operating at atmospheric pressure and 400 °C requires heat to be removed from the bed at the rate of 480 kW, by means of water cooled tubing immersed in the bed. Given the data in table 4.3, determine the necessary length of tubing which has to be immersed in the bed to meet this duty. The required surface area of tubing can be calculated from the well known equation for heat exchanger performance

$$\dot{Q} = UA_s \text{LMTD} \tag{4.24}$$

where $\dot{Q}$ is the duty or heat transfer rate, U is the heat transfer coefficient, A_s is the surface area required and LMTD is the logarithmic mean temperature difference between fluids.

Table 4.3 Data for Example 4.3.

Tubing bore	35 mm
Water inlet temperature	20 °C
Water outlet temperature	75 °C
Mean particle size	840 μm
Particle density	1540 kg m^{-3}
Fluidizing gas density	0.52 kg m^{-3}
viscosity	3.25×10^{-5} kg m^{-1} s^{-1}
thermal conductivity	5.1×10^{-5} kW m^{-1} K^{-1}
Minimum fluidizing velocity	0.19 m s^{-1}

Figure 4.6 shows the temperature distribution along the length of immersed tubing in relation to the bed temperature; the latter is regarded as being uniform through assuming good mixing of particles. The LMTD is given by

$$\text{LMTD} = \frac{\theta_1 - \theta_2}{\ln(\theta_1/\theta_2)} \tag{4.25}$$

where θ_1 and θ_2 are the temperature differences between the bed and liquid at inlet and outlet to the tube (see notation in figure 4.6). In this example

$$\text{LMTD} = \frac{(400-20)-(400-75)}{\ln[(400-20)/(400-75)]}$$

$$= 352 \text{ K}.$$

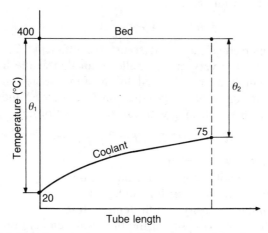

Figure 4.6 The temperature variation along a tube in a fluidized bed.

The thermal resistance of the tube wall may be neglected in this case so that the overall heat transfer coefficient U is given by

$$\frac{1}{U} = \frac{1}{h_b} + \frac{1}{h_w} \quad (4.26)$$

where h_b and h_w are the bed-to-immersed-surface heat transfer coefficient and that at the liquid–tube wall interface respectively. In this example we may calculate the maximum value of h_b using Zabrodsky's correlation given in equation (3.19), and then take 70% of this calculated value for h_b, after first checking that the Reynolds number, Re_{mf}, at minimum fluidizing velocity, equation (2.41), is less than 12.5 and the Archimedes number, Ar, equation (2.44), is less than 26 000, as shown in Example 3.2. From the data given in table 4.3

$$Re_{mf} = \frac{0.52 \times 0.19 \times (840 \times 10^{-6})}{(3.25 \times 10^{-5})}$$

$$= 2.55$$

$$Ar = \frac{0.52 \times (1540 - 0.52) \times 9.81 \times (840 \times 10^{-6})^3}{(3.25 \times 10^{-5})^2}$$

$$= 4407.$$

Inserting the data from table 4.3 into Zabrodsky's equation, (3.19):

$$h_{max} = 35.8 \times (1540)^{0.2} \times (5.1 \times 10^{-2})^{0.6} \times (840 \times 10^{-6})^{-0.36}$$
$$= 334 \text{ W m}^{-2}\text{ K}^{-1}.$$

Taking 70% of this value gives

$$h_b = 234 \text{ W m}^{-2}\text{ K}^{-1}.$$

The bed-to-immersed-surface heat transfer coefficient, h_b, will be the controlling one as it is very much smaller than that at the liquid side of the tube, h_w. This may be checked by an appropriate correlation for liquid flowing through tubes, e.g. equation (4.27) (see Simonson (1975) or other standard texts, dealing with forced convection).

$$Nu = 0.027 \, Re^{0.8} Pr^{0.33} (\mu/\mu_{wall})^{0.14} \tag{4.27}$$

where Nu is the Nusselt number
Re is the Reynolds number for liquid flow
Pr is the Prandtl number for the liquid
μ is the liquid viscosity at bulk temperature
μ_{wall} is the liquid viscosity at the tube wall temperature.

For this Example it can be shown that the value of h_w will be about $10 \text{ kW m}^{-2}\text{ K}^{-1}$, so that using equation (4.26) to determine the overall heat transfer coefficient U, we obtain

$$U = [(1/234) + (1/10\,000)]^{-1}$$
$$= 229 \text{ W m}^{-2}\text{ K}^{-1}.$$

Hence, using equation (4.24) and putting $A_s = \pi \times$ (the tube diameter $\times$ tube length), the tube length required

$$= \frac{480 \times 10^3}{\pi \times 0.035 \times 229 \times 352}$$
$$= 54.2 \text{ m}.$$

Having established the amount of tubing required, the designer has next to consider whether there is sufficient space in the bed to fit the 54.2 m length of tubing in easily, say in the form of a single flat zig-zag pattern, as shown in figure 4.7(a), as rows of series connected elements, as shown in figure 4.7(b), or as a set of parallel tube rows with the ends of the tubes connected to headers, as shown in figure 4.7(c). The arrangement in figure 4.7(c) gives rise to a lower water velocity through the tubes and, hence, a smaller heat transfer coefficient, h_w, at the liquid–tube wall interface. The overall heat transfer coefficient U is reduced by this reduction in h_w, although the reduction will not normally be very large because h_w is large by comparison with the bed-to-immersed-surface heat transfer coefficient, h_b, in equation (4.26). A corresponding increase in the length of tubing required will arise.

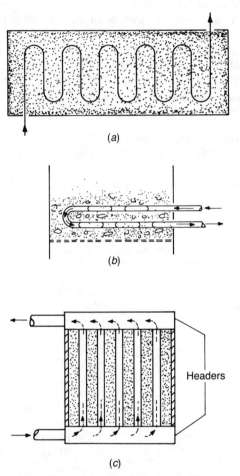

Figure 4.7 Arrangements of tubing in beds.

Whatever arrangement is chosen, the greater the length of tubing which has to be packed into the bed, the closer the adjacent lengths or convolutions, and this gives rise to interference with particle convection, with attendant reduction in the bed-to-immersed-surface heat transfer coefficient, h_b. The extent of this has to be established by experiments with the particular bed, but to give some sort of indication of magnitude, Elliott *et al* (1971) reported that with tube bundles having a gap between the tubes of 70 mm, the heat transfer coefficient was about

124 Design of Simple Fluidized Beds

10% lower than that with a single isolated tube and that reduction of the gap to about 25 mm led to about 25% of the values obtained with a single tube. If space is very limited, use of extended surface (i.e. finned) tubing might have to be considered.

We now pass on to a second example in which heat is removed by the fluidizing gas.

Example 4.4. 200 tonne/day of hot sand particles have to be cooled from 200 to 45 °C. Air is used as the fluidizing gas. Given the data in table 4.4, determine the size of the containment for the fluidized bed required to suit the following options:

(a) cooling entirely by fluidizing gas in a single fluidized bed;
(b) cooling entirely by fluidizing gas in a number of independent stages;
(c) cooling partly by fluidizing gas and partly by water cooled tubes installed in the bed.

Table 4.4 Properties of sand particles and fluidizing air.

Air Properties	
Pressure	1 atmosphere
Temperature at plenum	15 °C
Density at 15 °C	1.228 kg m^{-3}
at 45 °C	1.104 kg m^{-3}
at 143.5 °C	0.850 kg m^{-3}
at 169 °C	0.798 kg m^{-3}
Specific heat at constant pressure	1.005 kJ kg^{-1} K^{-1}
Viscosity at 15 °C	1.788×10^{-5} kg m^{-1} s^{-1}
at 45 °C	1.930×10^{-5} kg m^{-1} s^{-1}
at 169 °C	2.457×10^{-5} kg m^{-1} s^{-1}
Thermal conductivity at 45 °C	2.762×10^{-5} kW m^{-1} K^{-1}
at 122.5 °C	3.333×10^{-5} kW m^{-1} K^{-1}
Sand	
Mass flow rate	200 tonne h^{-1}
Mean particle size	0.520 mm
Particle density	2640 kg m^{-3}
Bulk density	1450 kg m^{-3}
Specific heat	0.8 kJ kg^{-1} K^{-1}
Inlet temperature	200 °C
Outlet temperature	45 °C
Minimum fluidizing velocity at 15 °C	0.213 m s^{-1}
at 45 °C	0.202 m s^{-1}
at 170 °C	0.166 m s^{-1}

Figure 4.8 depicts the simple fluidized bed into which the sand enters at 200 °C and leaves at 45 °C. In order to decide the volume of the containment vessel, a suitable value for the duration of time each sand particle should reside in the bed to cool it to the desired temperature has to be established. To do this we should first make an estimate of the time required for a single particle to be cooled from its initial temperature, 200 °C, to close to its final steady temperature, 45 °C. It will be recalled from Chapter 3 that the fluidizing gas at 15 °C reaches a bed temperature of 45 °C within a very short distance above the distributor; so it will be sufficient to estimate this time period using the assumption that a particle at 200 °C is cooled by a gas stream at 45 °C. This cooling time, sometimes known as the 'relaxation time', may be chosen somewhat arbitrarily to be that required for the particle to cool by 90% of the initial temperature difference (200–45) K. The particle residence time in the bed should then be at least ten times this 'relaxation time', so as to allow for a wide variation in both the residence time between individual particles and the fluidizing-gas-to-particle heat transfer coefficient.

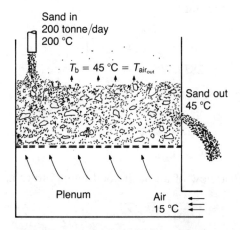

Figure 4.8 A fluidized bed sand cooler.

Appendix A shows the relaxation time for a single particle to be 7.94 s, under conditions applicable here. In the absence of experimental data giving a more accurate average and deviation of residence time, at least ten times this value should be allowed for; indeed, in practice, a residence time of two or three minutes is not uncommon. For the present purpose, allow two minutes.

The volume of sand in the bed, V_s, should therefore be

126 Design of Simple Fluidized Beds

$$V_s = \frac{\text{mass flow} \times \text{residence time}}{\text{bulk density}}$$

$$= \frac{(200 \times 1000)}{24 \times 60} \times \frac{2}{1450}$$

$$= 0.192 \text{ m}^3.$$

Now let us calculate the rate of heat removal from the sand and the air flow rate needed to do this. Remember that a considerable average residence time has been allowed for and that we assume that the mixing of gas and particles is very rapid, so that the bed is sensibly uniform in temperature; remember, also, from Example 3.1 that the air only has to penetrate a few millimetres into the bed before the particle and air temperature are almost identical. The fluidizing air may, therefore, be treated as leaving the bed at a bed temperature of 45 °C. Hence, since

$$\text{rate of heat removal from sand} = \text{rate of heat received by air} \quad (4.28)$$

$$\dot{m}_p C_{pp}(\Delta T_p) = \dot{m}_f C_{pf}(\Delta T_f) \quad (4.29)$$

where (ΔT_p) and (ΔT_f) are the temperature changes of particles and gas, respectively, the duty

$$= \frac{200 \times 1000 \times 0.8 \times (200 - 45)}{24 \times 3600}$$

$$= 287 \text{ kW}.$$

Hence, the mass flow of air

$$= \frac{287}{1.005 \times (45 - 15)}$$

$$= 9.52 \text{ kg s}^{-1}$$

and the volumetric flow of air at the bed temperature and pressure

$$= \frac{9.52}{1.104}$$

$$= 8.62 \text{ m}^3 \text{ s}^{-1}.$$

For a velocity of fluidizing gas of 0.5 m s^{-1}, this yields a planform area

$$= \frac{8.62}{0.5}$$

$$= 17.24 \text{ m}^2$$

and, since the volume is 0.192 m^3, the bed depth will be

$$= \frac{0.192}{17.24}$$

$$= 0.0111 \text{ m}$$

i.e. 11.1 mm. These bed dimensions lead to practical objections that a large floor area is required and that a small inclination of the distributor leads to a significant difference in bed depth between one end and the other (see figure 4.9), resulting in by-passing of the air through the shallowest part of the bed. (Non-uniform fluidization tends to impart the circulation of particles, similar to that encountered with a spouted bed mentioned in Chapter 1 and the whirling bed reported on by Rios *et al* (1980); this will not be discussed here however.) The choice of distributor, so as to avoid jets from the orifices punching right through the bed, also requires consideration. Further, as Example 3.1 shows, a few millimetres of bed depth above the distributor are required before the gas temperature approaches that of the particles; thus, with a bed 11.1 mm deep, the particle–gas mixing zone is not a small fraction of the bed.

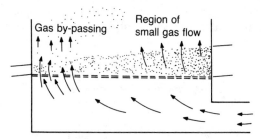

Figure 4.9 The effect of a small angle of tilt on the gas distribution in a very shallow fluidized bed.

On the other hand, suppose the most compact bed is desired, i.e. diameter = depth, then the fluidizing velocity will be excessive, in this case exceeding the entrainment velocity, thus

$$\text{bed diameter} = \left(\frac{4 \times 0.192}{\pi}\right)^{1/3} = 0.625 \text{ m}$$

giving a fluidizing velocity

$$= \frac{8.62}{(\pi/4) \times 0.625^2}$$

$$= 28.1 \text{ m s}^{-1}.$$

The way out of this dilemma is to alter the parameters; consider the following four possibilities.

(1) Increase the velocity of the fluidizing gas from 0.5 m s^{-1} (which is approximately $2.5 U_{mf}$) to a value below the elutriation limited velocity. This will reduce the plan area and increase the bed depth. Thus, suppose the gas velocity can be raised to 1.0 m s^{-1}, then the bed planform area will be 8.62 m² and the

128   Design of Simple Fluidized Beds

depth 22.2 mm, still somewhat shallow.
(2) Increase the bed depth further, but not so much as to give an excessive pressure drop and, hence, excessive fan pumping power. The reader may care to explore this possibility.
(3) Reduce the air flow so as to reduce the bed planform area and cool the sand in a number of independent stages, i.e. provide a staged fluidized bed cooler.
(4) Install water cooling tubes in the bed.

We will thus consider suggestions (3) and (4).

Staging of the beds

The fluidized beds may be arranged so that the solids flow in the opposite direction to that of the fluidizing gas flow, as shown in figure 4.10, this arrangement being termed 'counterflow'; or in 'crossflow', as shown in figure 4.11. The crossflow arrangement can meet the parameters of the particular sand cooling system concerned here more readily, as will be shown later.

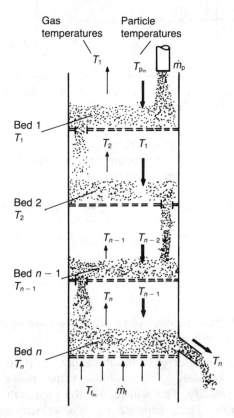

Figure 4.10 Fluidized beds arranged in counterflow.

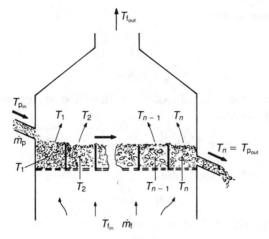

Figure 4.11 Fluidized beds arranged in crossflow.

Counterflow staging
Considering the *counterflow* arrangement in figure 4.10, and using the nomenclature thereon, it can be shown that for n beds, the steady flow energy equation for each bed is as follows:

Bed 1 $\quad \dot{m}_p C_{pp}(T_{p_{in}} - T_1) = \dot{m}_f C_{pf}(T_1 - T_2)$ (4.30)

Bed 2 $\quad \dot{m}_p C_{pp}(T_1 - T_2) = \dot{m}_f C_{pf}(T_2 - T_3)$ (4.31)

Bed n $\quad \dot{m}_p C_{pp}(T_{n-1} - T_n) = \dot{m}_f C_{pf}(T_n - T_{f_{in}})$. (4.32)

Eliminating the intermediate temperatures and putting the effectiveness of heat exchange for particles, η_p, as

$$\eta_p = \frac{T_{p_{in}} - T_n}{T_{p_{in}} - T_{f_{in}}} \quad (4.33)$$

and for gas, η_f, as

$$\eta_f = \frac{T_1 - T_{f_{in}}}{T_{p_{in}} - T_{f_{in}}} \quad (4.34)$$

and putting R = thermal capacity rate ratio

$$= \frac{\dot{m}_f C_{pf}}{\dot{m}_p C_{pp}} \quad (4.35)$$

yields

$$\eta_p = \left(\sum_{i=1}^{n} R^i\right)\left(1 + \sum_{i=1}^{n} R^i\right)^{-1} \quad (4.36)$$

$$\eta_f = \left(\sum_{i=1}^{n} R^i\right)\left[R\left(1 + \sum_{i=1}^{n} R^i\right)\right]^{-1}. \quad (4.37)$$

130 Design of Simple Fluidized Beds

Our requirement is that
$$\eta_p = \frac{200 - 45}{200 - 15}$$
$$= 0.838.$$

If the air flow rate is reduced such that five stages are employed, equation (4.36) becomes

$$\eta_p = \left(\sum_{i=1}^{5} R^i\right)\left(1 + \sum_{i=1}^{5} R^i\right)^{-1} = 0.838 \tag{4.38}$$

yielding a thermal capacity ratio
$$R = 1.01.$$

Note, here, that the air flow $\dot{m}_f$ is now
$$\dot{m}_f = \left(\frac{200 \times 1000}{24 \times 3600}\right) \times \frac{0.8}{1.005} \times 1.01$$
$$= 1.86 \text{ kg s}^{-1}.$$

It should be noted that the uppermost bed will have the highest temperature, its fluidizing velocity will be the greatest among the beds, but its minimum fluidizing velocity will be the lowest. This means that the fluidizing index U/U_{mf}, which may be regarded as a guide to the amount of elutriation, will be the greatest in this bed.

Correspondingly, the lowest bed will have the lowest fluidizing index U/U_{mf}. We should, therefore, choose a bed planform area to give air velocities and indices in both these beds which ensure that the elutriation limit is not exceeded in the upper bed and that minimum fluidization is exceeded adequately in the lowest bed.

First, the upper bed temperature T_1 may be found by noting from equations (4.36) and (4.37) that

$$\eta_f = \frac{\eta_p}{R} \tag{4.39}$$

so that
$$\eta_f = \frac{0.838}{1.01} = 0.83.$$

Then, using equation (4.34) with the inlet temperatures of the air and particles
$$T_1 = 0.83 \times (200 - 15) + 15 = 169 \text{ °C}.$$

Suppose that separate tests have shown that it is necessary to limit the fluidization index U/U_{mf} in the upper bed to about 5, then from the data on minimum fluidization in table 4.4, the air velocity in the upper bed will be $(5 \times 0.166) = 0.83$ m s^{-1}. Hence, the bed planform area, A_b, will be

$$A_b = \frac{1.86}{0.798 \times 0.83}$$

$$= 2.81 \text{ m}^2$$

while the air velocity, U_1, in the bottom bed, which is at a temperature of 45 °C, is

$$U_1 = \frac{1.86}{1.104 \times 2.81}$$

$$= 0.6 \text{ m s}^{-1}$$

which is about 2.97 U_{mf}. For the 2 min particle residence time previously suggested, yielding a total volume of particles of 0.192 m^3, the depth of each bed will be

$$= \frac{0.192}{5 \times 2.81}$$

$$= 0.0137 \text{ m}$$

i.e. 13.7 mm, which is still rather shallow. If a concession, permitting an increase in air velocity to, let us say, $10U_{mf}$ in the uppermost bed is made, then the bed depth will be doubled and, correspondingly, the bed planform area will be halved. However, the overall pressure drop remains the sum of five pressure drops across the distributors plus that across the five beds. Under these revised circumstances, allowing a pressure drop of 350 mm water gauge across each distributor yields an overall pressure drop of

$$5 \times \left(350 + \frac{1450}{1000} \times 2 \times 13.8\right)$$

$$= 1950 \text{ mm water gauge.}$$

It may be tempting to reduce the airflow rate further, but the scope for this is limited, since reducing the value of the heat capacity ratio, R, to 0.838 would, from consideration of equations (4.39) and (4.37), yield an effectiveness of heat transfer to the air of 100% and, thus, require an infinite number of stages.

Clearly, the counterflow arrangement has a high effectiveness of heat transfer to the cooling air and takes up less floor space; but it is of more complicated construction, has a higher pressure drop and, although the matter will not be gone into here, consideration has to be given to problems of controlling the flows of solids and gases within the cooler, e.g. the reliability of downcomers, clearance of blockages etc.

Crossflow staging

We consider very briefly the alternative arrangement of crossflow staging, as shown in figure 4.11. By writing the steady flow energy equations for each of the n equal sized beds having the same air flow through each stage, and remembering that the temperature of the air leaving the cooler will be the arithmetic mean of those leaving each stage, it can be shown (Levenspiel 1984) that the effectiveness of heat transfer to the particles and to the air is as follows:

$$\eta_p = \left(1 - \frac{1}{(1 + R/n)^n}\right) \tag{4.40}$$

$$\eta_f = \frac{1}{R}\left(1 - \frac{1}{(1 + R/n)^n}\right) \tag{4.41}$$

giving

$$\eta_f = \frac{\eta_p}{R}. \tag{4.42}$$

The required effectiveness of heat transfer to the solids η_p is 0.838. Using equation (4.40) leads to

$$(1 + R/n)^n = 6.17. \tag{4.43}$$

Suppose we have five stages, then

$$1 + R/5 = (6.17)^{1/5}$$

giving $R = 2.19$. The mass flow of air is then

$$= \frac{200 \times 1000}{24 \times 3600} \times \frac{0.8}{1.005} \times 2.19$$

$$= 4.04 \text{ kg s}^{-1}.$$

Using equation (4.42)

$$\eta_f = \frac{0.838}{2.196}$$

$$= 0.382.$$

The steady flow energy equation for the first bed (figure 4.11) is

$$\dot{m}_p C_{pp}(T_{p_{in}} - T_1) = \dot{m}_f C_{pf}(T_1 - T_{f_{in}}) \tag{4.44}$$

giving the temperature T_1 of the first bed

$$T_1 = \left(\frac{T_{p_{in}} + (R/n)T_{f_{in}}}{1 + R/n}\right) \tag{4.45}$$

where R is the heat capacity ratio, equation (4.35).

Inserting $n = 5$ and the temperatures given in table 4.4, the temperature T_1 of the first bed is then

$$T_1 = \frac{200 + (2.196/5) \times 15}{1 + 2.196/5}$$

$$= 143.5\,°\text{C}$$

at which the minimum fluidizing velocity is $0.17\,\text{m s}^{-1}$. Taking the fluidizing velocity through this bed as $5U_{mf}$ leads to the planform area $A_{b,1}$ as

$$A_{b,1} = \frac{4.04}{5} \times \frac{1}{0.850 \times 5 \times 0.17}$$

$$= 1.118\,\text{m}^2.$$

Hence, the total planform bed area = $5 \times 1.118 = 5.59\,\text{m}^2$.

The fluidizing velocity through the final stage will clearly be

$$= \frac{4.04}{5} \times \frac{1}{1.118 \times 1.104}$$

$$= 0.655\,\text{m s}^{-1}$$

which is $3.85\,U_{mf}$.

For the same volume of bed, $0.192\,\text{m}^3$, the bed depth will be

$$= \frac{0.192}{5.59}$$

$$= 0.0343\,\text{m}$$

i.e. 34.3 mm.

This is significantly deeper than those in the counterflow stages. Again, if a concession to allow the air velocity to be doubled is made, then the bed depth will be doubled and the planform area halved. With this concession, the pressure drop will still be significantly lower than with the counterflow staging. Allowing a 350 mm water gauge pressure drop across the distributor, the overall pressure drop will thus be

$$= \left(350 + \frac{1450}{1000} \times 2 \times 34.5\right)$$

$$= 450\,\text{mm water gauge}.$$

It will be left to the reader to experiment further with the effects of changing parameters, but it should be noted that if the thermal capacity ratio R is reduced too greatly it will not be possible to attain a solution to equations (4.40) to (4.42). For example our requirement is that the value of η_p will be 0.838. However, if the value of R is reduced to 1, the right-hand side of equation (4.40) can never equal 0.838, even with an infinite number of stages.

134 Design of Simple Fluidized Beds

We now turn to exploring the possible advantages of installing cooling tubes in the bed.

(c) Cooling tubes in the bed
Such an arrangement is shown in figure 4.12, in which the flow of sand is in counterflow to that of the cooling water flowing through the tubes. Part of the heat from the solids is carried away by the cooling water and part by the fluidizing air. The rate of heat removal from the particles, $\dot{Q}_p$, must be the sum of the rate of heat transfer to the water, $\dot{Q}_w$, and to the fluidizing air, $\dot{Q}_f$, thus

$$\dot{Q}_p = \dot{Q}_w + \dot{Q}_f. \tag{4.46}$$

Using the nomenclature in figure 4.12, the steady flow energy equation for the system is

$$\dot{m}_p C_{pp}(T_{p_{in}} - T_{p_{out}}) = \dot{m}_w C_{pw}(T_{w_{out}} - T_{w_{in}}) + m_f C_{pf}(T_{f_{out}} - T_{f_{in}}). \tag{4.47}$$

The rate of heat transfer to the cooling water, $\dot{Q}_w$, is given by

$$\dot{Q}_w = U_{ht} A_t \text{LMTD} \tag{4.48}$$

where U_{ht} is the overall heat transfer coefficient (bed-to-water) and A_t is the surface area of tubing.

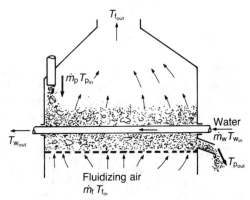

Figure 4.12 A water cooled fluidized bed sand cooler.

Equations (4.47) and (4.48) may be most conveniently solved by deciding an initial value for the fraction of the heat transfer from the solids which is to be carried away by the water, subject to certain practical limiting features.

(i) Clearly, if the entire heat transfer from the particles were to go to the cooling water then, according to equations (4.46) and

(4.47), no air flow would be needed. However, if there is no air flow there would be no particle convection and the value of the overall heat transfer coefficient would be extremely small; hence an enormous tube surface area would be needed.

(ii) The fluidizing velocity has to exceed U_{mf} everywhere and, in order to maximize the bed-to-immersed-surface heat transfer coefficient, the fluidizing velocity needs to be about twice that of U_{mf}.

(iii) The tube spacing and, if finned tubing is used, the fin spacing have to be sufficient to prevent too great a fall-off of heat transfer coefficient, yet be compact.

Well tested and reliable information, which would allow such considerations to be readily incorporated into a mathematical description of the heat exchanger, do not seem to be to hand. In order to make progress, however, suppose that the tubes have to remove 50% of the heat to be removed from the solids. Zabrodsky's equation for the maximum bed-to-immersed-surface heat transfer coefficient, h_{max}, for a single tube (see equation (3.19)) predicts $h_{max} = 342 \, W \, m^{-2} \, K^{-1}$ at a mean bed temperature of 122.5 °C. Taking 70% of this value, allowing a reduction of about 30% for deterioration due to close packing of the tubes and treating the heat transfer coefficient on the water side of the tubes as being large, the overall bed-to-tube-surface heat transfer coefficient will be about $0.2 \, kW \, m^{-2} \, K^{-1}$.

Suppose the water inlet and outlet temperatures to be 15°C and 80 °C, respectively, then from equation (4.25), the LMTD will be

$$\text{LMTD} = \frac{(200 - 80) - (45 - 15)}{\ln[(200 - 80)/(45 - 15)]}$$

$$= 64.9 \, K$$

and since the heat transfer rate from the solids found earlier is 287 kW, then from equation (4.48), the tube surface area, A_t, required is

$$A_t = \frac{(0.5 \times 287)}{0.199 \times 64.9}$$

$$= 11.1 \, m^2.$$

The bulk volume of particles in the bed has to be sufficient to accommodate this amount of tube surface. (Note that if a plain tube of 25 mm diameter were used, the total length of tubing would amount to 141 m.)

Taking the mean outlet temperature of the fluidizing air to be the arithmetic mean of the solids inlet and outlet temperatures, namely 122.5 °C, the mass air flow, $\dot{m}_f$, needed to remove 50% of the heat from the solids is

$$m_f = \frac{0.5 \times 287}{1.005 \times (122.5 - 15)}$$
$$= 1.33 \text{ kg s}^{-1}.$$

Taking the fluidizing velocity at the coolest end of the bed to be twice U_{mf} at that end, i.e. $2 \times 0.202 = 0.404 \text{ m s}^{-1}$, the planform area of the bed, A_b, will be

$$A_b = \frac{1.33}{1.104 \times 0.404}$$
$$= 2.98 \text{ m}^2.$$

Thus, the installation of cooling tubes in the bed, instead of staging the beds, has allowed a reduction in bed planform area from 5.56 m² in the case of the five-stage crossflow arrangement, and reduced the air flow demand.

It remains to fit the tubing into the bed without making the bed too deep and, thus, giving excessive pressure drop. This will not be attempted here, but to give some indication of possibilities, suppose the 141 m of plain 25 mm diameter tubing had to be fitted in. We may start by using 59 parallel tubes, each 2.39 m long, and laying them in three layers on a 63 mm triangular pitch in a box of planform dimensions 2.39 m × 1.25 m, as shown in figure 4.13, although the distance between adjacent tube surfaces is perhaps somewhat small. The bed depth required to accommodate this tube packing arrangement would be about 150 mm. One of the uncertainties in connection with the bed depth is the amount of bed expansion. Experiments would be required to verify that all tubes are sufficiently immersed.

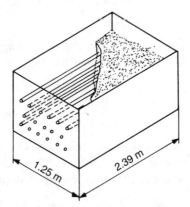

Figure 4.13 A schematic layout of tubing in a fluidized bed.

Further, it should be remembered that when a bank of tubes is installed in a fluidized bed, then the tubes block off part of the planform area of the bed and this results in a high gas velocity in the space between the tubes. Thus, for the tube packing arrangement proposed above, one row of 20 tubes of 25 mm outside diameter and 2.39 m length has a total projected area of $20 \times 0.025 \times 2.39 \text{ m}^2 = 1.2 \text{ m}^2$; thus, the gas velocity between the tubes is $[2.98/(2.98 - 1.2)] = 1.67$ times the 'empty bed' gas velocity which was used to estimate the bed size. Tests are required to assess the effect of this on performance and on elutriation.

Summary of Results of Example 4.4
Considerably more refinement would be needed to optimize the design of a fluidized bed cooler than has been shown in the above calculations. The results are, however, summarized in table 4.5. It should also be realized that recovery of heat from the sand, up to 287 kW, may be well worth while; the final column of the table shows some possible outputs.

4.6 Optimum Size of a Fluidized Bed Reactor

For simplicity of illustration, consider the problem of deciding suitable dimensions for a fluidized bed reactor, supplied with a prescribed rate of flow of fluidizing gas, which is necessary for the reactor to produce a prescribed rate of flow of product. The designer has to decide whether to build:

(a) a fluidized bed reactor vessel of large size, consuming a small amount of fan power to overcome the pressure drop across the bed, distributor and ducting, i.e. a high capital cost–low running cost option;

(b) a small fluidized bed reactor, requiring a high power input to the fan, i.e. low initial cost–high running cost;

(c) a reactor of intermediate size and fan input power.

Assuming all other aspects of these alternatives to be equal, the problem may be reduced to optimizing the capital cost of the reactor and its installation with the annual running cost, over the life of the plant. In practice, the latter should include costs of maintenance, but for simplicity, here, the cost of electrical energy may be regarded as the dominant item in running costs.

Criteria commonly used to assess the economics of such projects are based upon cash flow analysis and include:

(i) payback period;
(ii) internal rate of return on the capital.

Details of these and other aspects of the financing of projects may be found in standard texts, such as those listed in the Bibliography at the end of this chapter.

Table 4.5 Summary of results of calculations for a sand cooler (Example 4.4).

Arrangement	Planform area (m²)	Bed depth (mm)	Overall pressure drop (mm water gauge)	Heat from solids coverted to:
Single bed	17.2 or 8.6	11.1 22.2	— —	9.52 kg s^{-1} air at 45 °C
Five-stage beds: Counterflow	2.81 or 1.40	13.8 each 27.6 each	— 1950	1.86 kg s^{-1} air at 169 °C
Crossflow	4.36	44	450	4.04 kg s^{-1} air at 85.7 °C
Single stage with water tubes in bed	2.98	140	553	0.528 kg s^{-1} water at 80 °C plus 1.33 kg s^{-1} air at 122.5 °C

It may be helpful here however to take some fictitious data and work a simple example to illustrate the kind of approach involved in deciding the optimum design.

Example 4.5. Table 4.6 shows the estimated capital cost, running cost and annual net cash flow of different fluidized bed reactor designs that are intended to produce an identical quantity of saleable product annually from the identical inputs of solids and fluidizing gas. The independent variable is the fluidizing velocity at which the reactor operates. The annual revenue from sales of the product is 140 000 dollars and the life of the plant for purposes of this example is to be 5 years.

Table 4.6 Capital cost and net cash flow for a fictitious fluidized bed plant designed for different fluidizing velocities.

Design No	Fluidizing velocity (m s^{-1})	Capital cost (Dollars)	Annual running cost (Dollars/year)	Annual cash flow (Dollars/year)
1	0.5	200 000	52 800	87 200
2	0.6	171 000	58 400	81 600
3	0.8	128 000	71 900	68 100
4	1.0	98 000	84 500	55 500
5	1.1	89 500	91 700	48 300
6	1.2	85 600	111 000	42 300
7	1.4	79 800	123 000	29 000
8	1.6	72 200	136 000	17 000

Estimate the fluidizing velocity giving the shortest payback period and the largest internal rate of return over this 5 year period.

The payback period and internal rate of return have to be calculated for each of the designs and the results examined to see which of them produces:

(i) the shortest payback period;
(ii) the largest internal rate of return.

(i) *Payback period*
This is the number of years the reactor has to be in operation before the capital cost is equalled by the cumulative amount of cash flow.

For design No 1, the cumulative cash flow over the first two years is

$$-200\,000 + 87\,200 + 87\,200 = -25\,600 \text{ dollars}$$

while over the first three years it is

$$-200\,000 + 87\,200 + 87\,200 + 87\,200 = +61\,600 \text{ dollars.}$$

The payback period is thus between 2 and 3 years, namely 2.3 years.

A similar calculation for each design in turn has to be made. The results are plotted on a base of fluidizing velocity, as shown in figure 4.14 which shows that the shortest payback period will be given when the fluidizing velocity is about 1 m s^{-1}, and amounts to about 1.77 years.

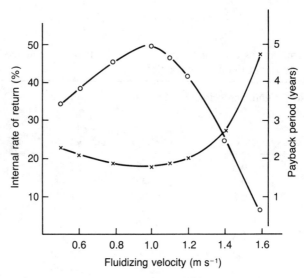

Figure 4.14 The internal rate of return (○) and payback period (×) as a function of fluidizing velocity from Example 4.5.

(ii) *Internal rate of return*

The internal rate of return, i, takes account of the time value of money by discounting the cash flow. It is estimated by putting the net present value (NPV) equal to zero and solving for i; thus for design No 1

$$\text{NPV} = -200\,000 + \frac{87\,200}{i}\left(\frac{(1+i)^5 - 1}{(1+i)^5}\right) \quad (4.49)$$

giving $i \simeq 33\%$.

Repeating this for each design in turn and plotting the internal rate of return on a base of fluidizing velocity (as in figure 4.14) shows that the greatest internal rate of return is about 49%, and occurs when the fluidizing velocity is about 1 m s^{-1}.

Comments
In this particular example, the use of both payback period and internal rate of return for the determination of the optimum fluidizing velocity yielded similar values. This may not always be the case; space does not however permit this issue to be pursued here and the reader is referred to standard texts for amplification.

More importantly, it has to be recognized that calculations such as those in Example 4.5 can sometimes be very sensitive to small changes in the input data. Analyses should therefore always include a sensitivity analysis and allowances for inaccuracies in data, plant reliability, etc should be made. The importance of obtaining reliable data, e.g. from specific commercial quotations, and including all cost items, can hardly be over-emphasized.

It is reiterated that the above example was a simplified one, so as to illustrate a basis for exploration of the effect of changes in one parameter, namely the fluidizing velocity, on financial criteria which might be used to judge the reactor design or, indeed, any project. In practice, an assessment may involve optimization of many more variables, but consideration of these variables is beyond the scope of this book.

4.7 Concluding Remarks

The purpose of this chapter has been to provide some elementary insight into considerations which are involved in the design of fluidized beds. Reactor modelling and scale-up of fluidized beds have not been discussed, nor have chemical reactions and rates entered into the arguments presented here. Mention has not been made of such problems as erosion. Such topics are to be found scattered throughout the literature, but the reader is referred to texts such as Yates (1983) and Davidson *et al* (1985); Grace (1974) has presented a concise paper dealing with the scope of such matters. Some brief comments on scale-up are also made in Chapter 6.

The worked examples have been concerned only with particles belonging to Geldart's group B. Space has not permitted consideration of particles belonging to group D or particles of group A. The former are coarse particles and are more difficult to fluidize; however, although they are encountered widely in practice, there is rather less data available about their behaviour. Fine particles, belonging to group A, are of such great importance in practice that a whole body of 'fine particle technology' has been developed.

Finally, it should be noted that the chapter has concentrated only on the bubbling regime. The turbulent and 'fast' regimes of fluidization, which are of increasing interest, are discussed briefly in Chapters 1 and 5.

References

Davidson J F, Clift R and Harrison D (ed.) 1985 *Fluidization* 2nd edn (London: Academic)
Elliott D E, Healey E M and Roberts A G 1971 *Proc. Conf. Inst. of Fuel and l'Inst. des Combustibles et de l'Energie, Paris, 1971*
Geldart D 1985 in *Fluidization* ed. J F Davidson, R Clift and D Harrison (London: Academic) ch 11
Grace J R 1974 Fluidization and its application to coal treatment and allied processes *AIChE Symp. Ser. No* 141 **70** 21–76
Levenspiel O 1984 *Engineering Flow and Heat Exchange* (New York: Plenum)
Rios G M, Baxerres J L and Gibert H 1980 in *Fluidization* ed. J R Grace and J M Matsen (New York: Plenum) pp 529–36
Simonson J R 1975 *Engineering Heat Transfer* (New York: Macmillan)
Yates J G 1983 *Fundamentals of Fluidized-bed Chemical Processes* (London: Butterworths)

Bibliography

Alfred A M and Evans J B 1971 *Discounted Cash Flow – Principles and Some Short Cut Techniques* 3rd edn (London: Chapman–Hall)
Archer S H, Choate G M and Racette G 1983 *Financial Management* 2nd edn (New York: Wiley)
Her Majesty's Stationery Office 1977 *Life Cycle Costing in the Management of Assets – A Practical Guide* (London; HMSO)
Leech D J 1982 *Economics and Financial Studies for Engineers* (Chichester: Ellis Horwood)
Ruegg R 1977 in *Waste Heat Management Guidebook, NBS Report* PB-264959 (Washington, DC: Federal Energy Administration) ch3
Stevens G T Jr 1979 *Economic and Financial Analysis of Capital Investments* (New York: Wiley)

5 Fluidized Bed Combustion

5.1 Introduction

5.1.1 Historical

The subject of the historical development of fluidized bed boilers and fluidized bed combustion has been written on by Skinner (1970) and Ehrlich (1976, 1984). Professor Arthur Squires (1983) and Squires *et al* (1985) have also traced developments. Most historical reviews, however, start from the accomplishments of Fritz Winkler who used fluidization techniques which resulted in a process to gasify lignite, in order to make fuel gas, at a plant near Leipzig, Germany in the 1920s.

The combustion of fuels in fluidized beds seems to have developed relatively rapidly since the 1960s, so that boiler and furnace manufacturers world-wide have included fluidized bed boilers and furnaces among their range of products for several years.

The author counts himself privileged indeed to have worked with (to quote Shelton Ehrlich) 'one of the heroes of the story', namely the late Professor Douglas Elliott who was one of those so instrumental in bringing the technology forward in the 1960s and 1970s.

5.2 Combustion Systems for Solid Fuels Generally

5.2.1 Inputs and outputs

Figure 5.1 represents, schematically, a combustion system enclosed within a system boundary. In ideal situations the fuel is a pure hydrocarbon and enters the system steadily together with a quantity of air slightly in excess of that required theoretically, so as to ensure complete combustion. The fuel burns completely and the combustion products, namely carbon dioxide (CO_2) and water vapour (H_2O), together with nitrogen (N_2) and the excess oxygen (O_2), leave the system in the exhaust gases, while heat, Q, is transferred from the system to the surroundings.

144 Fluidized Bed Combustion

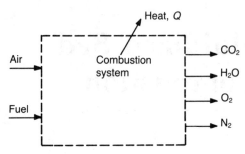

Figure 5.1 A thermodynamic representation of a combustion system.

However, all solid fuels are very complex compounds of carbon, hydrogen, sulphur, nitrogen and other inert and active components, e.g. chlorine, alkali metal salts, etc and moisture. In addition to CO_2, H_2O and O_2, the combustion of such fuels will be accompanied by the production of atmospheric pollutants (e.g. NO_x and SO_x) or corrosive, condensable vapours (e.g. H_2SO_4 or HCl). Further, the inert ash in the fuel does not vanish; part of it will be entrained as particulates in the exhaust gases, adding to the pollutants discharged, while part may remain in the system and may have to be removed mechanically.

This less-than-ideal situation is depicted by figure 5.2, illustrating also that not all of the fuel burns to completion. CO, unburnt hydrocarbon (HC) gases and unburnt carbon particles can also appear in the exhaust gases, so that the maximum possible amount of heat has not been liberated within the combustion system. Thus as well as being pollutants, CO, unburnt C and HC can also be designated 'efficiency reducers', along with the excess air supplied. In well designed and closely controlled boiler plant and furnaces fired by solid fuel, unburnt carbon particles and heat carried away by excess air are the biggest single contributors to loss of efficiency. Accordingly, considerable efforts are made by designers and plant operators to minimize both.

Nowadays environmental protection legislation in most countries sets limits to the concentration of pollutants which may be emitted from industrial plants, not least power stations, industrial boilers and furnaces. The particulates are separated from the gas stream using cyclones, electrostatic precipitators etc downstream of the combustion system. With conventional combustion equipment, SO_2 is removed most commonly by wet scrubbers (dry flue gas desulphurization techniques are also being developed). With fluidized bed combustion systems, however, the approach to prevention of excessive emission of SO_2 in the gases exhausted from the plant to the atmosphere is different (see §5.3).

It has always to be borne in mind, however, that the specification and choice of type of boiler or furnace plant has to meet the particular needs of the purchaser; this includes environmental and other requirements on site, capital and operating costs, flexibility of operation, etc (see Highley and Kaye 1983).

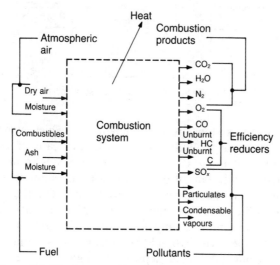

Figure 5.2 The inputs and outputs of a combustion system.

5.2.2 Combustors and the First Law of Thermodynamics

A combustor is intended to provide a source of heat by burning fuel with sufficient air to ensure efficient combustion. The heat liberated may be put to use for a variety of purposes, such as generating steam or hot water, providing a stream of hot gas (e.g. for use in drying processes), and the incineration of wastes.

It is necessary to distinguish between the heat which is utilized for an intended purpose, e.g. steam raising, and that which may have to be removed from the combustor to maintain acceptable working conditions within it. (In the latter case, cooling tubes or considerable quantities of air in excess of that required to effect complete combustion, will have to be provided to carry away the necessary quantity of heat. It is clearly desirable to put this removed heat to good use, but this may not always be done.)

A combustor of any type is a thermodynamic system and, accordingly, the First Law of Thermodynamics applies to it. Consider figure 5.3 which depicts a thermodynamic system within which chemical reactions, namely combustion reactions, occur. A system boundary is drawn around the combustor and mass, heat and work can be transferred across it, i.e. the combustor is enclosed within a 'control volume'. In the case of a combustor the work transfer is zero.

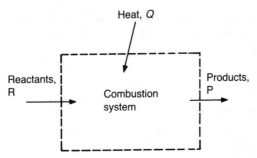

Figure 5.3 A combustion system.

The inputs to the system are the reactants, R, which would comprise fuel and air, but may include sorbents. In the case of conventional combustors, the exhaust gas scrubber to remove SO_2 from the exhaust gases is located downstream of the combustor and is not enclosed within the particular system boundary considered here. On the other hand, with fluidized bed combustors, flue gas desulphurization is achieved by feeding sorbent limestone or dolomite into the combustor (see §5.3) and, accordingly, the sorbent should be included in the input.

The outputs from the system depicted by figure 5.3 are the products P, comprising the normal products of combustion, namely CO_2, H_2O and excess O_2 and N_2, but also the pollutants SO_x, NO_x, ash and any sorbent transported out of the system. Heat, Q, also crosses the system boundary; the direction of heat flow shown in figure 5.3 is opposite to that with a combustor since, normally, heat is lost from the combustor to the surroundings, but the effect of this on subsequent computation is merely that heat loss from the combustor results in a negative quantity, as will be shown by a worked example later.

If conditions are steady and work transfer is zero, heat transfer Q to *the system from the surroundings* is related to the enthalpy, H†, of the reactants and products by the steady flow energy equation

$$H_R + Q = H_P \qquad (5.1)$$

where H_P is the enthalpy of the products and H_R is the enthalpy of the reactants.

The enthalpies H_P and H_R are dependent upon temperature above

†Enthalpy H is defined in standard texts on thermodynamics as

$$H = m(e + pv)$$

where m is the mass of the substance, e is the specific internal energy, p is the pressure and v is the specific volume (see, for example, Spalding and Cole 1973).

the datum temperature T_0, which is 25 °C, and the enthalpy of the particular reaction, $[\Delta H_0]$, at the above datum temperature (see standard texts, e.g. Spalding and Cole 1973). Values of $[\Delta H_0]$ for many reactions are to be found in tables and are mostly negative values, e.g. for the reaction of pure carbon with oxygen, $C + O_2 = CO_2$, the value is $-393\,500$ kJ kmol^{-1} of carbon. For solid fuels generally, however, the calorific value normally determined by the bomb calorimeter experiment is sufficiently close to the negative of $[\Delta H_0]$ for most combustion calculations. Note, also, that the reactants comprise fuel—including its ash and moisture content—and air—usually humid although often regarded as being dry.

Returning to figure 5.3, if the reactants enter the system at a temperature T_1, and the products leave at a temperature T_2, equation (5.1) may be rearranged and written as

$$Q = (H_{P_2} - H_{P_0}) + (H_{P_0} - H_{R_0}) + (H_{R_0} - H_{R_1}) \quad (5.2)$$
$$= (H_{P_2} - H_{P_0}) + [\Delta H_0] + (H_{R_0} - H_{R_1}) \quad (5.3)$$

where $[\Delta H_0] = (H_{P_0} - H_{R_0})$ is the enthalpy of reaction which, here, is approximately the higher calorific value of the fuel.

These quantities are illustrated in figure 5.4. Note, however, that where a sulphur sorbent is added with the fuel feed, it is a reactant and has its own enthalpy of reaction; these properties would have to be included in equations (5.1) to (5.3). However, for simplicity of example, consider Example 5.1 below where no sorbent is involved.

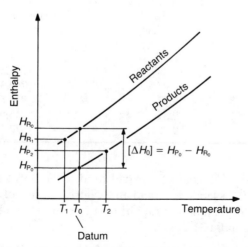

Figure 5.4 An enthalpy–temperature diagram for the combustion process.

Example 5.1. A combustor is enclosed in a chamber having cooled walls, as shown in figure 5.5. Air and fuel, which includes moisture and ash, enter the combustion chamber at a pressure of 1 atmosphere and a temperature of 15 °C, where they are burnt at constant pressure and discharged at 850 °C. The mass of each constituent in the exhaust gases per kg of fuel burnt to completion is given in table 5.1. The stoichiometric air : fuel ratio is 9.84 by mass and 20% excess air is supplied.

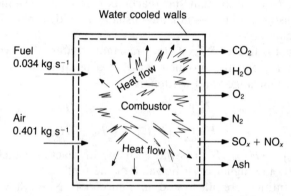

Figure 5.5 A combustor for Example 5.1.

Table 5.1 The mass and specific enthalpy of exhaust gas constituents.

Constituent	Mass in exhaust gases (kg/kg fuel)	Specific enthalpy at 850 °C relative to datum at 25 °C (kg/kg constituent)
CO_2	2.90	914.2
H_2O vapour	0.33	1777.5
$SO_x + NO_x$	0.05	760.3
N_2	9.27	912.5
O_2	0.51	846.0
Ash	0.08	577.5

Taking the specific heat of the fuel fired to be $0.94 \text{ kJ kg}^{-1} \text{ K}^{-1}$, that of the ash as $0.7 \text{ kJ kg}^{-1} \text{ K}^{-1}$ and the calorific value of the fuel to be $29\,650 \text{ kJ kg}^{-1}$, determine the rate of heat loss to the combustor walls for a fuel flow rate of 0.034 kg s^{-1}.

We use equation (5.3) to determine the heat loss Q per kg of fuel to the combustor walls, taking $T_2 = 850\,°C$, $T_1 = 15\,°C$ and $[H_0] = 29\,650 \text{ kg/kg}$ of fuel.

Products

The term $[H_{P_2} - H_{P_0}]$ is the sum of the enthalpies of the constituents above datum, namely

$$[H_{P_2} - H_{P_0}] = m_{CO_2}h_{CO_2} + m_{N_2}h_{N_2} + m_{O_2}h_{O_2} + m_{SO_x + NO_x}h_{SO_x + NO_x}$$
$$+ \text{[enthalpy change of the } H_2O \text{ in cooling from}$$
$$850\ °C \text{ to } 25\ °C]. \quad (5.4)$$

Note that the H_2O will be entirely in the vapour phase at 850 °C, but at 25 °C a fraction of it will be in the liquid phase and the remainder in the vapour phase. Thus

$$\text{enthalpy change of } H_2O = m_{vapour}h_{vapour} \text{ (at 850 °C)}$$
$$- m_{liquid}h_{liquid\ to\ vapour} \text{ (at 25 °C)}. \quad (5.5)$$

In this case, the mass of H_2O in the liquid phase at 25 °C will be 0.07 kg, so that

$$\text{enthalpy change of } H_2O = 0.33 \times 1777.5 - 0.07 \times (-2442)$$
$$= 757.5\ kJ/kg \text{ of fuel.}$$

The results of the computation of the enthalpy of the products (equation (5.4)) are given in table 5.2, together with the summation.

Table 5.2 Computation of the enthalpy of the products.

Constituent	Enthalpy at 850 °C relative to datum at 25 °C (kJ/kg of fuel)	
CO_2	2.9×914.2	= 2651.2
H_2O	$0.33 \times 1777.7 - 0.07 \times (-2442)$	= 757.5
$SO_x + NO_x$	0.05×760.3	= 38.0
N_2	9.27×912.5	= 8458.9
O_2	0.51×846.0	= 431.5
Ash	0.08×577.5	= 46.2
	Summation = $[H_{P_2} - H_{P_0}] = 12\,383.3$ kJ/kg fuel	

Reactants

$$[H_{R_0} - H_{R_1}] = (m_{fuel}c_{p,\ fuel} + m_{air}c_{p,\ air})(T_0 - T_1)$$
$$= (1 \times 0.94 + 1.2 \times 9.84 \times 1.005)(25 - 15)$$
$$= 128.1\ kJ/kg \text{ fuel.}$$

Inserting these values into equation (5.3) gives the required heat flow per kg of fuel:

$$Q = 12\,383.3 + (-29\,650) + 128.1$$
$$= -17\,138.6 \text{ kJ/kg fuel}.$$

The negative sign means heat removal.

For a fuel flow rate of 0.034 kg s^{-1}, the required heat flow rate $\dot{Q}$ to the combustor walls is then

$$\dot{Q} = 17\,138.6 \times 0.034$$
$$= 583 \text{ kW}.$$

5.3 Fluidized Bed Combustion of Solid Fuels

5.3.1 General remarks

Figure 5.6 shows a schematic diagram of a fluidized bed combustion system. In the simplest arrangement, the bed consists of inert particles such as sand, through which fluidizing air can be passed via a distributor plate at the bottom. When operating normally, the bed contains only a small percentage of burning fuel, perhaps 1 or 2%, so that the inert

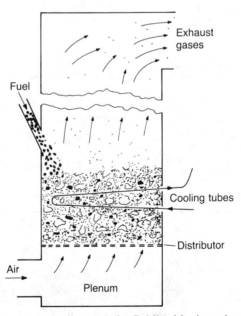

Figure 5.6 A schematic diagram of a fluidized bed combustor having in-bed cooling surfaces.

particles transport the heat away from the burning particles, thus keeping their temperature below that at which ash melts.

Before the bed can burn solid fuel, its temperature has to be sufficiently high for solid fuel to ignite readily when it is fed into the bed. A start-up system is thus needed; the various types of start-up system will not be discussed here, however, but Moodie and Vickers (1985) give an account of start-up systems in current use. The bed temperature at which fuel feeding commences varies greatly with the type of fuel being burnt and has to be determined from trials; for coals 500 °C is not uncommon. Dependent upon the type of fuel and operating conditions, some combustion may take place in the zone above the free surface of the bed as well as in the bed.

A system to feed fuel to the bed and deal with the ash is also required. Fuel may be fed from above the bed or injected beneath the surface. Provision has to be made to separate entrained ash from the exhaust gases and to remove ash accumulated in the bed.

The temperature of the bed when burning solid fuels is commonly controlled to be about 850 °C; this may necessitate heat removal from the bed by installing cooling tubes, as indicated in figure 5.6, or by other means. The bed temperature and heat removal are commented on further in §§3.2 and 4.2.

Fluidized bed combustion (FBC) systems are capable of burning solid, liquid and gaseous fuels with low emissions of pollutants; most FBC systems designed and built, however, have been done so for the burning of solid fuels.

Although FBC systems offer considerable advantages, they are not a panacea and each application has to be thought out properly. It may be wise to recall the professional gambler's prescription, 'horses for courses'! Fluidized bed combustion is likely to be the favourite horse for the course for example, when low-grade, sulphurous fuel has to be burnt on a site where stringent limits on emission of pollutants are imposed. In the converse situation other 'runners' may be preferred.

Some favourable features of fluidized bed combustion systems are listed and commented upon in the next section.

5.3.2 Combustion temperature

The combustion temperature is controlled at a relatively low value of about 850 °C and this has a number of advantages.

(i) It is below the melting or softening temperature of ash, so that the ash does not stick to the heat transfer surfaces or tubes and, hence, form deposits on tubes which reduce their heat transfer capability.

(ii) The ash formed is seldom hard, eliminating one of the substances responsible for erosion.

(iii) The temperature is lower than that required to vaporize alkali metal salts found in the ash (thereby eliminating corrosive vapours) and lower than the melting point of vanadium and similarly troublesome metals found in some fuels.

5.3.3 Control of pollutant emission

With fluidized bed combustion systems, flue gas desulphurization is brought about by adding a sorbent such as limestone or dolomite to the fluidized bed, where the sulphur is absorbed in a solid form. Stantan (1983), among others, gives a detailed account of this technique. In some countries the environmental regulations demand removal of 90% of the sulphur dioxide. A combustion temperature of about 850 °C is about the optimum for absorption of sulphur oxides by the lime, and the residence time of the lime in the bed is lengthy. A scrubber in the exhaust gas duct is thus not needed, because little of the SO_x reaches the exhaust gas. Flue gas desulphurization is not achieved free of charge however, irrespective of whether the combustion system is a fluidized bed or any other type—you cannot have something for nothing! Sorbent has to be supplied in significant quantities; this has consequences for the amount of solids to be handled by the plant, both at input and output; think about the extra particulate loading in the exhaust gases, sorbent disposal, etc. Experience with fluidized bed combustion systems shows that the amount of sorbent required in order to absorb about 90% of the sulphur in the fuel is that giving a calcium : sulphur molar ratio of about 3. Work aimed at obtaining improved sorbent utilization continues.

The emission of oxides of nitrogen is a complex study and the reader is referred to the literature on the subject (see for example, Shaw 1983). It has been demonstrated that, in some circumstances, nitrogen in the fuel is the source of most of the NO_x emitted. A widely advocated approach for reducing NO_x emissions is to modify the bed to operate in a two-stage mode (Cooke 1984, Pentland and Mitchell 1984). Meantime developments continue.

5.3.4 Size of combustion chamber and tolerance to changes of fuel quality

Fluidized bed systems are generally smaller for a given thermal output than conventional equipment. The extent to which this occurs will depend upon circumstances; detailed comparisons are beyond the scope of what can be dealt with here, but as a rough guide, the firing rate ($MW\,m^{-2}$) of a bubbling bed fluidized bed boiler having cooling tubes in the bed can be up to 50% greater. Furthermore, fluidized bed combustion systems are more tolerant to changes in fuel quality or grade, i.e. they are less selective in the type of fuel they can burn and can tolerate wider fluctuations in fuel quality. Low-grade fuels may be burnt more readily in fluidized beds than in other types. Indeed with some low-grade fuels and wastes, e.g. colliery tailings, fluidized bed combustion is about the only way in which they can be burned satisfactorily.

The relatively good mixing of particles and gas promoted by bubbling action provides good conditions for both combustion and heat transfer. As pointed out in Chapter 3, the bed-to-immersed-surface heat transfer coefficient is several times larger than values met in forced convection conditions at atmospheric pressure, resulting in smaller surface area requirements for heat removal

5.3.5 Pressurized fluidized bed combustion

If a coal fired fluidized bed combustor is operated at elevated pressure, the products of combustion can be expanded through a gas turbine to produce mechanical work and hence generate electricity. The products of combustion have to be sufficiently clean for a gas turbine to accept without excessive erosion, corrosion or fouling of the turbine. Accordingly, the necessary gas cleaning has been the subject of much research. Pressurized fluidized bed combustion (PFBC) has been under extensive research and development, not only because of its potential as a coal fired gas turbine, but also when used in conjunction with a steam plant (see figures 5.7 and 5.8 for alternative systems). A significantly higher efficiency of electricity generation is possible than that from

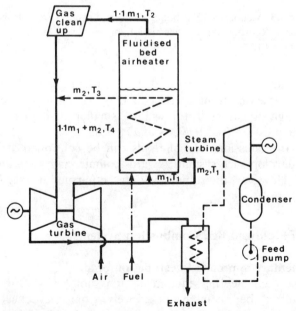

Figure 5.7 A basic 'air' cycle or air heater combined cycle. (From Roberts *et al* 1983.) Reproduced by permission of Elsevier Applied Science Publishers Ltd.

either a gas turbine or steam turbine plant alone (see Roberts *et al* 1983, Evans and Anastasiou 1985).

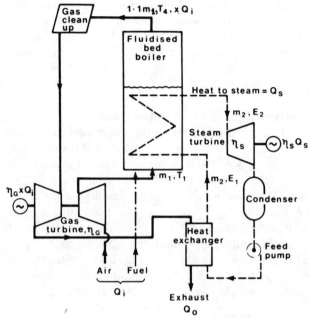

Figure 5.8 A supercharged boiler combined cycle. (From Roberts *et al* 1983.) Reproduced by permission of Elsevier Applied Science Publishers Ltd.

Clearly, the size of a pressurized fluidized bed combustor for a given rate of heat release (from a given fuel, operating at a given air : fuel ratio, at a given gas velocity) will be smaller than one operated at atmospheric pressure (see Roberts *et al* 1983).

Further, the emissions of pollutants can be controlled at low levels. Efforts to develop such systems towards commercialization are currently continuing. However, the subject will not be pursued further here.

5.4 Size of Fluidized Bed Combustion Systems

5.4.1 Elementary approach to estimation of size

As with most engineering equipment, it is not realistic to attempt to design fluidized bed combustors entirely from theoretical concepts. Competitive commercial designs are the product of fundamental understanding, acquisition of data, much practical experience with the equipment (which includes learning from mistakes) and proper economic assessment. None of this is acquired cheaply!

An excellent review of a considerable variety of designs of fluidized bed combustion systems for boilers and furnaces operating in the bubbling regime has been made by Highley and Kaye (1983), while the proceedings of the many international conferences on fluidized bed combustion quoted in the References at the end of the chapter contain further detail. It will be sufficient for the purposes of this chapter to note that, at the time of writing, many fluidized bed combustion systems operating in the bubbling regime have approximate values of some design parameters, as shown in table 5.3.

Table 5.3 Approximate values of some design parameters for fluidized bed boilers and furnaces (bubbling regime).

Fluidized bed boilers
Firing rate when burning coals:
  about 2 MW m^{-2} bed planform area when cooling tubes are installed in the bed
  about 1 MW m^{-2} with no tubes in the bed

Fluidizing velocity:
  1–3 m s^{-1}

Bed temperature:
  normally about 850 °C, but at minimum load about 800 °C

Turn down, i.e. (maximum)/(minimum) thermal output:
  about 2.5 for a single bed—multiple beds required for larger turn down

---

Furnaces
Firing rates when burning coals are about 75% of those for the above. Fluidizing velocities and turn down are similar to those above, but bed temperatures are about 950 °C and, at minimum load, about 800 °C.

Clearly there are insufficient data and information from which to design a fluidized bed boiler or furnace. It may, however, be instructive to make a rough assessment of sizes and heat transfer requirements, in order to illustrate some of the problems for which the designer has to find solutions.

Example 5.2. A fluidized bed combustor is required to operate at atmospheric pressure and a bed temperature of 850 °C. The fuel and excess air are those used in Example 5.1. The density of air at 850 °C may be taken as 0.3145 kg m^{-3}.

Estimate the bed planform area required on the alternative bases (*a*)

156 Fluidized Bed Combustion

and (b) below for a fuelling rate of 5 MW:

(a) firing rate = 2 MW m^{-2};
(b) fluidizing velocity = 2.5 m s^{-1}.

Basis (a)

$$\text{Planform area} = (\text{fuelling rate})/(\text{firing rate}) \qquad (5.6)$$

$$= 5/2.0$$

$$= 2.5 \text{ m}^2.$$

Basis (b)

The data in Example 5.1 show the higher calorific value to be 29 650 kJ kg^{-1} and for 20% excess air the air : fuel ratio is 1.2 × 9.84 = 11.81. Hence, mass flow of fuel

$$\dot{m}_f = \frac{5 \times 1000}{29\,650}$$

$$= 0.169 \text{ kg s}^{-1}.$$

The mass air flow

$$= 11.81 \times 0.169$$

$$= 1.996 \text{ kg s}^{-1}.$$

With an air density of 0.3145 kg m^{-3} and a velocity of 2.5 m s^{-1}, the planform area

$$= \frac{1.996}{0.3145 \times 2.5}$$

$$= 2.54 \text{ m}^2.$$

Thus, bases (a) and (b) yield similar results. It is clear, however, that if a lower air velocity had been chosen, a larger planform area would have been required. The *number and location of fuel feed points* and type of feeder required to ensure reasonably uniform fuel distribution also require consideration because of imperfect lateral mixing.

5.4.2 Heat removal requirements

Example 5.3. Estimate the rate of heat removal required from the fluidized bed combustor used for Example 5.2.

Example 5.1 showed that, irrespective of the type of combustor, if the particular fuel was burnt to completion with 20% excess air, and the exhaust gas exit temperature was 850 °C, 17 138.6 kJ of heat had to be removed for every kg of fuel burnt. Thus, for a 5 MW combustor, the required rate of heat removal

$$\dot{Q} = 17\,138.6 \times 0.169$$
$$= 2900 \text{ kW}$$

i.e. 2.9 MW. The problem, now, is how to remove heat from the fluidized bed combustor at this rate. The most obvious things to do are to install cooling tubes in the fluidized bed and/or to provide water cooled containment walls. The depth of bed required will depend upon the arrangement of heat transfer surfaces, e.g. how much heat is taken out through the bed walls and distributor, how much heat is radiated from the free surface of the combustor to the above-bed zone, how much in-bed tubing has to be accommodated (see figure 5.9 for illustration) and, further (see Stantan 1983), whether or not sulphur retention is required. For simplicity here, however, let us ignore sulphur absorption considerations. An estimate of the amount of surface area required has to be made using principles outlined in Chapters 3 and 4, using the appropriate value of heat transfer coefficient for the location of each surface. The latter presents some difficulty in view of variations in published data. However, let us illustrate the problem further by the next numerical example, accepting that the values of heat transfer coefficient and free surface radiation may be subject to significant error.

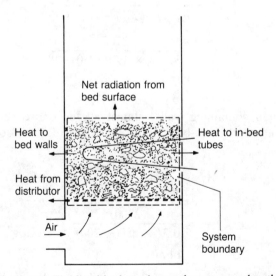

Figure 5.9 Fluidized bed combustor heat removal paths.

Example 5.4. Estimate the depth of bed required for the 5 MW fluidized bed combustor in Example 5.3 on the following alternative bases:

158   Fluidized Bed Combustion

(a) all heat release takes place in the bed, no cooling tubes in the bed, net radiation from the free surface of 200 kW, water cooled containing walls at a temperature of 90 °C, bed-to-wall heat transfer coefficient of 0.3 kW m^{-2} K^{-1};

(b) as (a) but with insulated bed containing walls and cooling tubes in the bed, a tube temperature of 200 °C, bed-to-immersed-tube heat transfer coefficient of 0.3 kW m^{-2} K^{-1}.

Rate of heat removal through walls

$\dot{Q}_w$ = total rate required − rate radiated from bed surface

= 2900 − 200

= 2700 kW.

Basis (a)

If the bed planform area of 2.5 m² is assumed to be circular, the diameter D is

$$D = \left(\frac{4 \times 2.5}{\pi}\right)^{1/2}$$

= 1.78 m.

The bed depth

$$H = \frac{\dot{Q}_w}{h_{b-w}\pi D(T_{bed} - T_{wall})} \qquad (5.7)$$

$$= \frac{2700}{0.3 \times \pi \times 1.78\,(850 - 90)}$$

= 2.12 m.

Basis (b)

With insulated containing walls, almost all the entire heat removal, $\dot{Q}_w$, must be effected by in-bed tubes, hence the tube surface area, A_t, required is

$$A_t = \frac{\dot{Q}_w}{h_{b-t}(T_{bed} - T_{tube})} \qquad (5.8)$$

$$= \frac{2700}{0.3 \times (850 - 200)}$$

≃ 14 m².

If this surface area is made up of 75 mm tubing, a length of 59.6 m is required. This could be accommodated in a bank of tubes comprising five horizontal rows of pipes having ten pipes of average length 1.2 m at 150 mm triangular pitch; the bank would require a bed depth of less than 0.5 m.

Thus the in-bed cooling tubes give a shallower bed, with the virtues of smaller bed pressure drop.

In practice, a combination of alternatives (*a*) and (*b*) in the above example could be used to obtain an even shallower bed. It must be stressed here again that the values of the parameters used here, e.g. the value of the heat transfer coefficients and the assumption that they are the same at the wall as at a tube immersed in the bed, and the dimensions arising from the above calculation, do not represent a refined design; they merely provide a first insight into the influence of heat removal considerations on the bed size. The improved mathematical modelling required for commercially acceptable design is outside our scope at this point.

Heat may also be removed from the bed by providing sufficient excess air, although this leads to loss of efficiency of boiler plant. Some types of fluidized bed boilers do not have cooling surfaces installed in the bed (see Highley and Kaye (1983) where other techniques for the necessary heat removal are also discussed). With fluidized bed furnaces or hot-gas generators used for dryers the required output is simply a stream of high-temperature gases. Depending upon the use which is to be made of the hot gases, they can be the combustion products from the furnace or, alternatively, clean air heated from air cooled tubes located in the bed and above-bed zone (again, see Highley and Kaye 1983).

5.4.3 Size of inert particles in the bed

The function of the inert particles in a fluidized bed combustor is to carry the heat away from burning fuel particles and to deliver it to a place where it can be utilized; e.g. to heat transfer surfaces where steam can be generated, to the excess air, or to the free surface where heat will be radiated.

The particles have to be capable of withstanding bed operating temperatures readily, be resistant to thermal shock, to attrition, and be reasonably cheap. Silica sand or alumina are commonly used. A suitable mean particle size and size range for each application has had to be established by a considerable amount of experimentation. The following points should be noted.

(i) If the particles are small, the bed-to-surface heat transfer coefficient will be high (see equation (3.19)) and the heat transfer surface area required to extract heat at a given rate will be minimized.

(ii) However, the terminal velocity of small particles is low and the fluidizing velocity at which the bed can operate without excessive elutriation is accordingly low. Hence, for a given fuelling rate at a given air : fuel ratio, a large bed planform area is needed. Conversely so for large particles.

A compromise between minimizing the amount of heat transfer surface area and minimizing the planform area of the bed has to be struck, but the incentive to maximize throughput, i.e. firing rates, has led to particle sizes in excess of 1 mm being used.

5.4.4 Velocity of fluidizing gas

At the most elementary level of consideration, the air velocity must exceed the minimum fluidizing velocity U_{mf} considerably, in order to achieve a high firing rate, yet be less than that which gives excessive elutriation of unburnt carbon particles, with its attendant loss of combustion efficiency. It is safest to base a prediction of combustion efficiency and carbon loss upon experimental results with a pilot plant, rather than theoretical analysis. Loss of combustion efficiency due to escape of unburnt carbon can be combatted by increasing the above-bed height of the combustor and the amount of excess air and bed depth, but the latter increases the fan pumping power. Recycling the elutriated material back to the bed can also be employed to combat carbon loss; however, the heat capacity of the recycled inert material alters the heat balance of the bed and the distribution of heat transfer surfaces throughout the system should reflect this fact. Deciding the most appropriate value for the air velocity is thus a complex matter; at the time of writing air velocities in the bubbling-bed-type fluidized bed combustors are within the range $1-3 \text{ m s}^{-1}$.

Significantly higher air velocities are employed in circulating fluidized bed combustors, where a considerable amount of solids are recycled; these types of fluidized bed combustor operate at considerably higher velocities where the fluidization regime lies between that of the bubbling bed and pneumatic transport (Reh *et al* 1980, Nack *et al* 1980, Stockdale and Stonebridge 1985, Basu 1986). This type of combustor will be discussed later.

5.4.5 Turn down

Combustors must be capable of operating stably at a fraction of their continuously rated output. The turn-down ratio (TDR) is simply

$$\text{TDR} = \frac{\text{maximum output}}{\text{minimum output}}. \quad (5.9)$$

For a single bed the TDR is commonly about 2; for a much larger turn down, separate beds operated in parallel are needed. Thus two beds, each having a TDR of 2, give a turn down equal to 4; however, control of the plant is more complex. We consider briefly the problem of how to achieve a practicable TDR for a single bed.

Consider the fluidized bed combustor explored in Example 5.4. If the fuel flow rate is halved and the air flow reduced by half in order to maintain the same percentage excess air, then the total rate of heat removal required from the system will be halved, i.e. to 1450 kW from 2900 kW. This requires that the amount of tube surface area immersed in the bed has to be reduced. The amount required can be estimated from equation (5.8), inserting (1450 − 200 free surface radiation =) 1250 kW for Q_w; this yields the required area as being about 6.5 m², instead of about 14 m² at full rating.

In principle, a change of tube surface area in contact with the bed occurs because of the relationship between bed expansion and air flow (see figure 5.10) and this feature may be exploited. However, the extent to which the tube surface area actually immersed in the bed changes with operating conditions has to be investigated by experiment, to see if it matches the turn-down requirement. Further, as indicated in figure 5.10, the heat transfer coefficient in the 'splash zone' above the bed surface is smaller than in the bed because of the lower concentration of particles per unit volume there, and it varies with operating and design parameters. The location of the tubes is thus important and the contribution of the splash zone to heat transfer needs to be taken into account in design calculations (see, for example, data by Chakraborty and Vickers (1984) and the model discussed by Moodie and Vickers (1985) in relation to start-up systems).

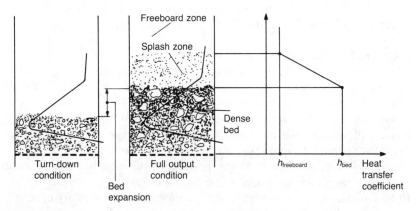

Figure 5.10 Regions of a fluidized bed combustor showing changes in the immersed length of cooling tube with bed expansion.

A smaller but useful addition to the TDR can be achieved if the bed temperature is allowed to fall, although 800 °C is about the minimum allowable temperature for reasons of combustion efficiency. Referring to equation (5.8) and neglecting any change in bed-to-immersed-surface heat transfer coefficient, the insertion of 800 °C instead of 850 °C for the new tube area of about 6.5 m² gives a heat removal rate by the tubing of 1150 kW. This amount and the amount of free surface radiation are reduced somewhat by a reduction in the radiative components of heat transfer due to the lower bed temperature and the area of particles seen at the reduced bubbling activity; the computation of these components is outside our consideration here. It is left as an exercise for the reader to estimate the effect of reduced bed temperature on turn down by making a thermodynamic analysis of heat removal requirements along the lines of Example 5.1, but with products leaving at 800 °C.

5.4.6 Fluidized bed combustors without in-bed tubes
Fluidized bed furnaces and some types of fluidized bed boilers do not have in-bed tubes. Nevertheless, the First Law of Thermodynamics still holds good, so that the steady flow energy equation (5.1) and its derivatives, equations (5.2) and (5.3), have to be satisfied. No worked example will be provided here, but it will be sufficient to say that excess air has to do the bed cooling instead of in-bed cooling tubes or water cooled walls; the reader can verify this by application of equation (5.3). Similarly, if the fuel is low grade, i.e. of low calorific value and high ash content, use of equation (5.3) along the lines of Example 5.1 will show that the heat removal requirement is smaller.

Other methods of achieving the essential heat balance, such as recycling fines, two-stage combustion etc can be found in the list of conferences in the References at the end of the chapter.

5.4.7 Distribution of heat release zones
The calculations of heat removal requirements made in Examples 5.1, 5.3 and 5.4 assumed complete combustion of the fuel within a defined system boundary (the bed alone in Example 5.4). However, different fuels burn in different ways. For example, solid fuels having a large content of volatile matter evolve the volatiles rapidly as the fuel particle becomes heated. A significant fraction of these volatiles will burn in the above-bed zone and some of the fine char may also burn there, while the larger residual char burns in the bed until its size is sufficiently small to be elutriated. On the other hand, some fuels are relatively unreactive and contain a much smaller amount of volatiles; such fuels often require a high bed temperature for satisfactory combustion.

Thus, the distribution of heat release zones and the rates of heat release within them depend upon the fuel characteristics and the situation is more complicated than the cases analyzed in the above simple examples. Space does not permit a more detailed modelling of fluidized bed combustors here. The wide range of types of fuel and their differing behaviour should always be borne in mind when designing a fluidized bed combustion system, because they affect decisions on the distribution of heat transfer surfaces, the employment of secondary air, sulphur capture, the type of fluidized bed combustor to be employed, etc.

5.5 Efficiency of Fluidized Bed Combustion Equipment

5.5.1 Efficiencies

Various efficiencies can be defined for use in assessing the performance of boilers or furnaces, and each has advantages and disadvantages. Generally (see Fenton 1977) for large boilers having high efficiency, a more accurate estimate of efficiency results if the losses are determined and the efficiency, η, calculated from

$$\eta = \frac{\text{input} - \text{losses}}{\text{input}}. \qquad (5.10)$$

The losses referred to in equation (5.10) are the sum of the enthalpies arising from:

(a) moisture in the fuel
(b) H_2O in the flue gases resulting from combustion of hydrogen in the fuel
(c) dry flue gases
(d) ash in the flue gases
(e) incomplete combustion of combustible gases
(f) incomplete combustion of carbon
(g) radiation and other unaccounteds for heat transfer.

Disregarding (a) (over which a designer may have little control and variations of which will have to be allowed for in design of the plant) and (g) which is often small, the above losses may be placed into two categories, namely those due to insufficient cooling of the flue gases (i.e. (b) to (d)) and those due to incomplete combustion ((e) and (f)).

Figure 5.11 is a generalized graph (from figure 5 in Highley and Kaye 1983) which shows how the principal losses from a typical coal fired fluidized bed boiler can vary with the amount of excess air supplied. The actual value of the efficiency obtained, however, is very dependent upon the design of boiler. The important points to notice from figure 5.11 are that (i) the largest component of the losses is that due to the elevated temperature of the flue gases and (ii) unburnt carbon is invariably the largest component of those losses due to incomplete combustion of the fuel. Theoretically, (i) could be reduced significantly if the flue gases were cooled to a lower temperature, but this is not done in practice because of the risk of serious corrosion problems arising from operating below the acid dew point of the gases and the cost of extra surface area required if sub-acid-dew-point operation is envisaged.

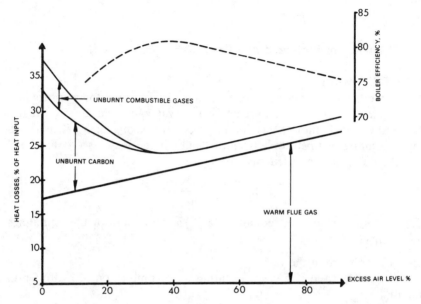

Figure 5.11 The effect of excess air on boiler heat losses of a typical coal fired boiler. (From Highley and Kaye 1983.) Reproduced by permission of Elsevier Applied Science Publishers Ltd.

Attention will therefore be focused now upon the efficiency of carbon burn-up.

Figure 5.12 shows a schematic diagram of a fluidized bed combustor, fired with solid fuel and fitted with a cyclone and bag filter for separating all the entrained particulates from the flue gases. Assume that all ash is captured by the gas cleaning equipment.

For the evaluation of carbon burn-up we define the 'combustion efficiency', η_c, for combustion of a solid fuel such as coal as

$$\eta_c = \frac{\text{(energy input from the fuel)} - \text{(energy in the unburnt carbon)}}{\text{energy input from the fuel}} \quad (5.11)$$

which may be written as

$$\eta_c = \frac{m_f(\text{HCV}_f) - m_C(\text{CV}_C)}{m_f(\text{HCV}_f)} \quad (5.12)$$

where m_f is the mass of fuel supplied, m_C is the mass of carbon unburnt, (HCV_f) is the higher calorific value of the fuel and (CV_C) is the calorific value of carbon.

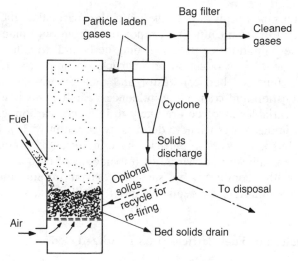

Figure 5.12 A fluidized bed combustor with exhaust cleaning equipment and grit re-firing.

The mass m_C of carbon unburnt has to be determined by collecting the mass of solids carried over while a mass of fuel m_f is burnt, analyzing these solids for carbon content and, hence, determining the mass of carbon unburnt for a given mass of fuel.

Clearly, if the solids captured by the gas cleaning equipment were to be recycled back to the bed, as shown in figure 5.12, then the unburnt carbon in the solids would be burnt further and the combustion efficiency increased. It is largely a matter of economics as to whether recycling is done. Recycling of solids back to the bed is not to be confused with operating in the 'fast' fluidization regime described in §5.9. Research has shown that the mechanisms of combustion of carbon are complicated and that interaction between chemical reactions at the carbon surface, heat and mass transfer, and burning particle temperature have to be elucidated further than at present. It has to be stated that, as yet, there is no way of *predicting* from available equations the mass of unburnt carbon which will be carried out of a given fluidized bed combustor under specified conditions with sufficient accuracy or discrimination for most purposes, e.g. to guarantee that a combustion efficiency of, say, 99% will be achieved. This is because the physical parameters within the bed, the burning particle temperature, the mass transfer coefficient and the chemical rate coefficient, needed for insertion in the equations, are not known with sufficient accuracy (see La Nauze 1985) and because elutriation rates and the size distribution of elutriated particles cannot be predicted sufficiently accurately. The magnitude of carbon 'carry over' from a given fluidized bed combustor can be determined only by trials with the particular fuel on the particular plant or from cautious extrapolations from pilot plant trials.

Despite the limitations of theory, it may be helpful to consider events from the moment of introduction of a particle or lump of solid fuel into the bed of a fluidized bed combustor and then discuss mechanisms of carbon combustion and controlling influences. Firstly, this is in order to see how the variables involved are likely to influence carbon 'carry over' and hence combustion efficiency, defined by equation (5.12). Secondly, rates of combustion of carbon particles have important consequences for control of the combustor, by virtue of the amount of carbon required to reside within the combustor to sustain a given output under given conditions. Such matters are dealt with briefly in the next section.

5.6 Combustion of Fuel Particles in a Fluidized Bed

5.6.1 Sequence of events

Consider figure 5.13, which shows the sequence of events and temperature history of a single particle of solid fuel, initially at ambient temperature, following it being introduced into a fluidized bed of inert particles. The air particles are maintained at a high temperature, typically 850 °C, and sufficient excess air is provided for satisfactory combustion.

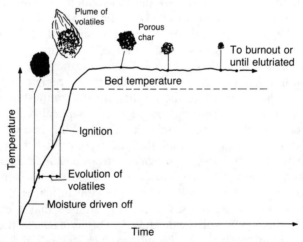

Figure 5.13 The temperature history of a single fuel particle introduced into a fluidized bed combustor.

Initially, moisture will be driven off and the temperature of the solid fuel particle rises as it is heated by inert particle convection, gas convection and radiation. The temperature of the fuel particle rapidly reaches a value at which volatile matter in the fuel commences to be liberated, escaping in the form of combustible vapours into the bed. The residue left after devolatilization is char, which is essentially porous carbon with some ash bound in it; the degree of porosity depends upon the type of fuel. Once the ignition temperature is reached, the char commences to burn, but, as indicated in figure 5.13, the time required to burn it out is much longer than that taken to evolve the volatiles.

It should be remembered that, unlike simple evaporation of a substance without chemical change, volatiles are the products of endothermic chemical reactions which occur on heating of the fuel. These products are combustible hydrocarbon gases which contribute significantly towards the calorific value of the fuel. (Solid fuel properties listed in Rose and Cooper (1977) show anthracites, bituminous coals and lignites having volatiles contents of about 5–10%, 30% and exceeding 50%, respectively.)

It is desirable to burn the volatiles in the bed, but in most cases a fraction of them may burn in the above-bed zone. If there is sufficient excess air and good mixing of volatiles and oxygen in the above-bed zone, and the temperature and mixture residence time are also sufficiently high, these gases and vapours will burn before being exhausted from the combustor. Naturally, the designer has to ensure that emissions of unburnt hydrocarbon gases or carbon monoxide are kept very small indeed. It must, however, be pointed out (see, for example, Turnbull and Davidson 1984) that the combustion of volatiles in fluidized bed combustors is not well understood at present.

The escape of volatiles from the solid fuel leaves the residual char porous, as indicated in the upper part of figure 5.13. As combustion of this char proceeds, the mass and size of the char particle(s) diminish until, eventually, these particles are sufficiently small and light to be elutriated from the bed. The elutriated char particles will continue to burn in the freeboard zone, diminishing in size further, as illustrated in figure 5.14. The smaller the size and mass to which each char particle burns down to within the combustor, the smaller the amount of unburnt carbon and the higher the combustion efficiency (equation (5.12)). Not all types of solid fuel remain in one piece when introduced into a hot fluidized bed, as suggested in the upper part of figure 5.13. With some solid fuels, the particles break into a number of smaller fragments as immediately as they are introduced into the fluidized bed (see Peçanha and Gibbs 1984). The number of elutriable size particles of carbon produced from each particle of fuel is correspondingly large and thus the combustion efficiency when burning such fuels may suffer. Other complications such as attrition can occur (see D'Amore *et al* 1980, Donsi *et al* 1981), which causes the production of large numbers of fine particles of fuel which can burn out rapidly or be elutriated. Burning within the pores of the char increases the rate at which the mass of the char particle diminishes, since a greater area than the superficial outer surface area of the particle is reacting at any moment. Such aspects render the mathematical description of char burning difficult and limited but, in combination with experimental work, a valuable insight into controlling mechanisms has been gained.

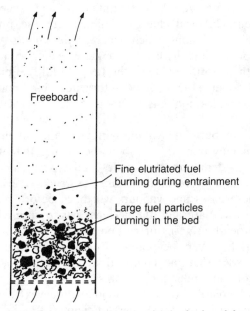

Figure 5.14 Char particles burning in the bed and freeboard.

The rate at which a carbon particle burns is of great importance and this subject is addressed briefly next.

5.6.2 Burning rate of char particles
5.6.2.1 Significance of burning rate
The 'burning rate' $\dot{m}_C$ requires definition. The rate of loss of mass of the particle is termed the 'burning rate' and may be expressed in kg s^{-1}, g s^{-1}, etc. The 'specific burning rate' $\dot{S}_C$ is the rate of mass loss per unit surface area and may be expressed in kg s^{-1} m^{-2} etc. Thus, for a spherical particle

$$\dot{S}_C = \frac{\dot{m}_C}{\pi d^2}. \tag{5.13}$$

Ideally, each char particle should reside within the combustor for sufficient time for it to burn completely. The faster the burning rate, the shorter the time required for the char particle to burn out. If, for simplicity, we consider the history of a single large char particle, as shown in figure 5.14, it will reside in the bed until it has become sufficiently small for it to be elutriated. Once elutriated, the char will continue to burn in the freeboard zone, but it is likely to be swept out of this zone rapidly so that burnout may not necessarily be completed and the combustion efficiency (see equation (5.12)) be less than 100%.

If the combustion efficiency is unacceptably low, the incompletely burnt particle may then have to be re-introduced into the combustor and burnt down further, or other measures to increase the rate of burn-up of fines be taken in order to achieve the combustion efficiency desired. This leads to the need to understand something about the mechanisms of combustion involved.

5.6.2.2 Mechanisms of combustion of carbon

The pioneering paper by Avedesian and Davidson (1973) laid the foundation for understanding and subsequent work by others has elucidated the mechanisms, controlling the combustion of carbon particles in fluidized beds further. An account of research findings is given by La Nauze (1985).

We now discuss some factors influencing burning rates in an elementary way.

A char particle whose surface is at a sufficiently high temperature will burn rapidly if there is a large mass flow rate of oxygen towards it. On the other hand, if the carbon surface is at a relatively low temperature, or the char is relatively unreactive, it may burn slowly, even though a copious supply of oxygen is transferred towards the char surface.

There are thus two resistances which impede the rate at which the mass of char reacts, namely that to mass transfer of oxygen, R_g, and that due to chemical kinetics, R_c, at the surface; these are illustrated diagrammatically in figure 5.15. To speed up the burning rate of the char, these resistances must be reduced. With a given char particle, the resistance to mass transfer of oxygen may be reduced by forced convection, while that due to chemical kinetics may be reduced by increasing the temperature of the char surface.

Mass transfer rate of oxygen

The mass transfer rate of oxygen towards the surface may be characterized by the Sherwood Number, Sh, given by

$$Sh = \frac{k_g d_c}{D_g} \quad (5.14)$$

where k_g is the mass transfer coefficient, d_c is the diameter of the char

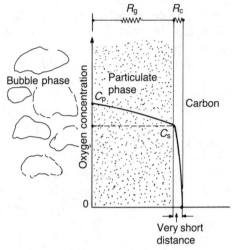

Figure 5.15 Resistances impeding the burning rate. R_g is the mass flow resistance and R_c is the surface chemical resistance.

particle and D_g is the diffusivity of oxygen. The Sherwood Number Sh is related to the carbon particle Reynolds Number Re ($= (\rho_f U d_c / \mu_f)$) and the Schmidt Number Sc ($= (\mu_f / \rho_f D_g)$) by correlations of the form

$$Sh = 2(1 + cRe^{1/2}Sc^{1/3}) \tag{5.15}$$

where c lies between 0.3 and 0.35.

Alternatively, Chakraborty and Howard (1981a) suggested that the voidage ε be taken into account by including it in the expression to give

$$Sh = 2\varepsilon + 0.69Re^{1/2}Sc^{1/3}. \tag{5.16}$$

Thus, to increase the Sherwood number and, hence, the mass transfer rate of oxygen, we must increase the particle Reynolds number and, in a fluidized bed, that means increasing the fluidizing velocity.

The specific burning rate $\dot{S}_C$ of the carbon corresponding to the mass flow rate of oxygen may be written as

$$\dot{S}_C = \lambda k_g (c_p - c_s) \tag{5.17}$$

where λ depends upon which particular oxidation reaction occurs at the surface (see equations (5.18) and (5.19) below), c_p is the oxygen concentration in the particulate phase of the bed (i.e., remote from the burning particle) and c_s is the oxygen concentration at the burning surface.

$$C + O_2 \rightleftharpoons CO_2 \tag{5.18}$$

for which $\lambda = \frac{3}{8}$.

$$C + \tfrac{1}{2}O_2 \rightleftharpoons CO \tag{5.19}$$

for which $\lambda = \tfrac{3}{4}$. Note here that the mass transfer coefficient, k_g, is the inverse of the resistance, R_g, to mass flow of oxygen:

$$R_g = 1/k_g. \tag{5.20}$$

Chemical kinetic rate

For a first-order reaction, the specific burning rate $\dot{S}_C$ may be expressed as

$$\dot{S}_C = k_c c_s \tag{5.21}$$

where k_c is the reaction rate coefficient and c_s is the oxygen concentration at the carbon surface.

The reaction rate coefficient k_c is normally expressed in an Arrhenius form:

$$k_c = A \exp(-E/RT_s) \tag{5.22}$$

where A is a frequency factor, E is the activation energy of the carbon, R is an ideal gas characteristic constant and T_s is the absolute temperature of the carbon surface.

Again, note the inverse relationship between the reaction rate coefficient and the surface chemical kinetic resistance R_c:

$$R_c = 1/k_c. \tag{5.23}$$

Thus, if chemical kinetics are the dominant controlling influence, the burning rate may be increased by increasing the surface temperature of the carbon.

Combination of mass transfer and chemical kinetic factors

The oxygen concentration at the char surface, c_s, is not known directly, so if it is eliminated between equations (5.17) and (5.21), the resulting expression for specific burning rate $\dot{S}_C$ becomes

$$\dot{S}_C = K c_p \tag{5.24}$$

where

$$1/K = 1/\lambda k_g + 1/k_c. \tag{5.25}$$

If the mass transfer coefficient, k_g, in equation (5.25), is replaced in terms of the Sherwood number

$$Sh = (k_g d_c / D_g) \tag{5.26}$$

for which correlations exist (see equations (5.15) and (5.16)), equation (5.25) may be written as

$$1/K = d_c / \lambda Sh D_g + 1/k_c. \tag{5.27}$$

We now have an equation, (5.24), with which to predict the specific burning rate, $\dot{S}_C$, and, hence, the burning rate, $\dot{m}_C$, of a solid carbon particle (i.e. a shrinking sphere model) when the appropriate values of the parameters for insertion into equations (5.27) and (5.22) are known. These are seldom known with sufficient accuracy, a prediction of the surface temperature of carbon being one of the most uncertain since it depends upon the heat balance for the char particle.

This heat balance is difficult to quantify because of the difficulty of predicting how much of the heat liberated in the reaction zone surrounding the burning particle reaches the char surface (Turnbull and Davidson 1984). It has been shown, however (see Ross and Davidson 1981, Chakraborty and Howard 1978), that the surface temperatures of carbon range from 25 to 250 K above that of the bed.

It should be reiterated that the above treatment is elementary. The resistance to mass flow of oxygen to the carbon particle will be mainly that due to the particulate phase of the bed, but any ash layer which may adhere to the particle also impedes mass flow; the empirical correlations, equations (5.15) and (5.16), do not distinguish between these resistances. Further, no proper account has been taken of char porosity. The oxygen concentration within the different phases of the bed is also uncertain; Minchener *et al* (1984) used a zirconia probe to measure this, but the subject needs further exploration. During its life in a fluidized bed, a burning char particle is alternately in the particulate phase and bubble phase of the bed, so that the resistances are fluctuating; experimental results, however, only express the net effect.

It is clear that during the residence time of a burning carbon particle within a fluidized bed, both chemical kinetic and mass flow rate resistances influence the burning rate. For example, over their range of experiments, Ross and Davidson (1981) showed that for particles of size 1–3 mm, combustion is mainly controlled by diffusion of oxygen to the surface of the carbon, whereas for particles smaller than 1 mm, combustion is controlled primarily by chemical kinetics. It was observed that the larger char particles glowed more brightly than the bulk of the bed, whereas the finer char particles glowed with about the same brilliance as the bed. This means that with the smaller char particles, the temperature of the char was controlling the burning rate and, since their temperature was only slightly above that of the bed, it would seem that the only way to speed up the burning rate of the *fines* is to raise the bed temperature.

Some of the fine char will still be elutriated from the bed before it has burnt out, however, and will continue to burn, probably under kinetic control, as it passes through the above-bed zone. The net effect of raising the bed temperature will be to reduce the carbon 'carry-over' and hence raise the combustion efficiency. The ash fusion temperature, however, will limit the extent to which raising the bed temperature can be exploited.

Generally, as reported by Chakraborty and Howard (1981b), the carbon burnup rate is promoted by using the largest practicable bed inerts, the highest fluidizing velocities and the highest bed temperatures.

Research into burning rates continues (see Prins and van Swaaj 1986) and each international conference reports further studies.

5.6.3 Minimum concentration of carbon to sustain combustion
5.6.3.1 Thermal stability
For simplicity, consider an element of a fluidized bed burning carbon particles, so as to avoid having to consider combustion of volatiles. For the thermal equilibrium of this element, the rate of heat liberation within it due to combustion of the carbon particles must equal the rate of heat removal. The heat removal may arise from a combination of surfaces immersed in the bed, heat losses through the bed containment and free surface radiation; all of these depend on the bed temperature, T_b, and the temperature of the surface, T_s, receiving the heat. The whole situation is described by

$$\dot{m}_a q \eta = h_{bs} A_s (T_b - T_s) + A_{fs} E (T_b^4 - T_z^4) + \dot{m}_a C_{pa} (T_b - T_0) \quad (5.28)$$

where q is the heat liberated per unit mass of air when all oxygen is utilized
η is the fraction of oxygen in the air which is consumed by combustion in the bed
$\dot{m}_a$ is the mass flow of air supplied to the bed
h_{bs} is the bed-to-surface heat transfer coefficient
A_s is the surface area immersed in the bed
T_b is the bed temperature
T_s is the immersed surface temperature
A_{fs} is the free surface area
E is the radiation exchange factor
T_z is the temperature of surfaces receiving radiation
C_{pa} is the specific heat capacity of air
T_0 is the temperature of air entering the bed.

Thus, to sustain the bed temperature T_b, the fraction of oxygen supplied to the bed by the air, η, which is completely converted into combustion products, must be sufficiently large. Clearly, the limiting temperature of the element of the bed, T_b, is the ignition temperature of the carbon, T_{ig}, otherwise combustion would cease. Hence, T_{ig} may be substituted for T_b in equation (5.28).

The fraction of the oxygen supply which can be consumed, η, (i.e. the conversion) depends upon the effectiveness with which the oxygen contacts the surface of the carbon particles and how rapid the carbon–oxygen reaction is. The former depends upon the bed behaviour, the operating parameters and the amount of active carbon surface area present; for a discussion on methods of predicting the magnitude of conversions, the reader is referred to Grace (1986). It will be sufficient here to state that, among other parameters, the air velocity and bed depth influence the amount of carbon in any element of the bed required to sustain combustion. Figures 5.16 and 5.17 (from Basu 1976) show some specimen numerical predictions.

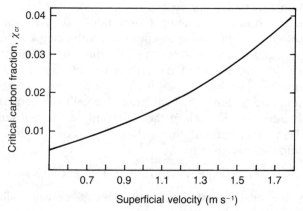

Figure 5.16 The variation of the critical carbon fraction with different velocities in a bed 0.25 m high and operating at 800 °C. $U_{mf} = 0.23$ s^{-1}, $d_i = 3$ mm, $H = 0.2$ kW m^{-2} k^{-1}. (Redrawn from Basu 1976.) Reproduced by permission of P Basu.

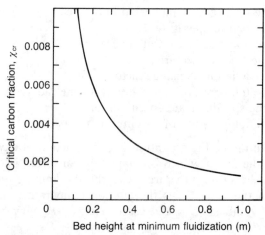

Figure 5.17 The variation of critical carbon fraction with bed height while the velocity is only 0.5 m s^{-1}. $d_i = 3$ mm, $T_b = 800$ °C. (Redrawn from Basu 1976.) Reproduced by permission of P Basu.

5.6.3.2 Effects of lateral mixing of solids in the bed

If a fluidized bed combustor is to rely on the natural mixing processes of the bed to distribute fuel particles throughout it then, in the case of a large bed, the thermal stability of the bed will dictate the spacing of fuel feed points. This arises because lateral (horizontal) mixing is significantly less vigorous than vertical mixing; consequently, fuel particle concentration gradients can arise in the lateral direction. This subject and its consequences were explored by Highley and Merrick (1971) and are illustrated by the simple diagram of figure 5.18. A lateral non-uniformity of fuel particle concentration leads to an excessive fuel concentration in the vicinity of fuel feeding points with an attendant high elutriation of fuel particles, a loss of combustion efficiency, and a reducing atmosphere near the feed point with an oxidizing atmosphere in other regions. In the extreme case of the fuel concentration being allowed to fall below the critical value discussed in §5.6.3.1, combustion will cease locally. Nozzle spacing on a large bed is also governed by economics and Highley and Merrick's work led to the recommendation that fuel feed nozzles should be spaced at intervals not exceeding about 0.9 m.

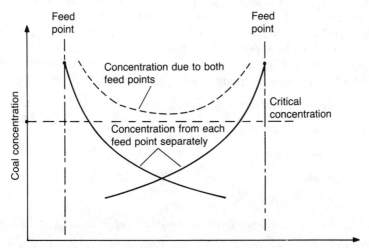

Figure 5.18 The distribution of coal concentration between feed points.

5.6.4 Volatiles

5.6.4.1 Influence on combustor design

It has already been mentioned that, although relatively little is understood about the combustion of the volatile matter content of solid fuels in fluidized bed combustors (§5.6.1), volatiles can contribute significantly to the calorific value of the fuel and influence combustor performance.

176   Fluidized Bed Combustion

If, for example, a combustor were designed for a solid fuel of very low volatiles content, and the fuel feeding and elutriation conditions were such that all combustion and, hence, heat release were completed in the bed, then all the heat transfer surface required for heat removal from the combustion zone would be located in the bed. Conversely, if the fuel has a high volatiles content and the bed depth and other features were such that most of the volatile matter burnt in the above-bed zone, leaving only sufficient residual carbon behind in the bed to keep the bed thermally stable, then the installation of cooling surfaces in the bed would be detrimental. In such a case, the provision of heat transfer surfaces in the above-bed zone would still be needed, in order to remove heat, but an appropriate distribution of them throughout the system requires consideration. The following simplified example indicates the sort of situation.

Example 5.5. A fluidized bed combustor burns solid fuel of calorific value $25\,000\text{ kJ kg}^{-1}$, the volatiles content being high and combustion conditions such that 40% of the calorific value is released in the above-bed zone and the remainder in the bed. The bed temperature is 850 °C, the air inlet temperature to the bed is 25 °C and the air : fuel ratio by mass is 13 : 1. Taking the specific heat of the products leaving the free surface of the bed as 1.04 kJ kg^{-1}, (a) estimate the amount of heat which has to be removed from the bed per kg fuel and (b) estimate the amount of heat per kg fuel to be removed from that section of the above-bed zone up to the point where the products leave at 850 °C.

The situation is as illustrated in figure 5.19.

(a) Heat removed from the bed, Q_b

$$= \text{heat released} - (\text{heat carried away by the products}) \quad (5.29)$$
$$= 0.6 \times 25\,000 - 14 \times 1.04\,(850 - 25)$$
$$= 2988\text{ kJ/kg fuel}.$$

(b) Heat to be removed from above-bed zone to cool products to 850 °C

$$= 0.4 \times 25\,000$$
$$= 10\,000\text{ kJ/kg fuel}.$$

Although the above calculation lacks precision and is sensitive to values of the parameters chosen, it illustrates nonetheless that where a significant amount of heat release occurs in the above-bed zone, most of the heat which has to be removed for the exit temperature of the products to be 850 °C has to be removed from the above-bed zone rather than the bed.

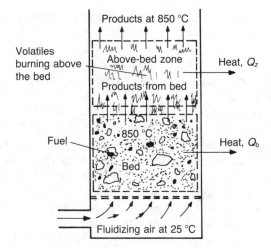

Figure 5.19 The diagram for Example 5.5.

Some care has to be taken about quantifying the amount of surface required in the above-bed zone. On the one hand, too little surface would result in excessive temperatures at the gas exit and within the above-bed zone, with attendant risks of fouling of heat transfer surfaces by softened ash. On the other hand, too much heat transfer surface may chill the flame, slowing down the reactions, and thus reducing the combustion efficiency and causing soot deposits. Despite the desirability of matching the heat transfer surface area distribution to the pattern of heat release zones, the matter has received scant attention to date.

A further important aspect concerning volatiles which are released in the bed and burn in the above-bed zone is whether the air flow should be split into two streams, the primary part passing through the bed and the secondary part, perhaps pre-heated, being supplied direct to the above-bed zone. Clearly, the smaller air flow through the bed would allow a smaller plan area of the bed or a lower gas velocity, the latter being important if there is excessive elutriation of carbon from the bed. A discussion of the engineering of such a system is, however, beyond the scope of this book.

Before leaving the subject of provision of secondary air, it should be noted that carrying out combustion in two stages is also employed as a technique for reducing the emission of oxides of nitrogen, and this has been the overriding reason for its implementation.

5.6.4.2 Effect of fuel particle size on rate of release of volatiles
Pillai (1985) carried out tests with two different types of coal, to establish the time required for the volatile matter to be released after immersion in a fluidized bed and how this release time was varied with

178   Fluidized Bed Combustion

the size of the coal particle. He found that the devolatilization time, t_v, followed a law of the form

$$t_v = kd_i \tag{5.30}$$

where d_i is the initial size of the coal particle and k is a constant for the coal and conditions.

Among the conclusions drawn was that the devolatilization times of large coal particles are as long as—and probably much longer than—typical solids mixing times in fluidized beds. This has importance for the distribution of heat release and temperature, particularly when *small* particles are being fed to the bed and transverse mixing is considered, in the following way. Figure 5.20 shows conceptually the transverse distribution of released volatiles for the cases of large coal particles and small ones. It will be seen that with the small particles the volatiles will have been released so rapidly that a large fraction of them may have burnt before the particles have had time to migrate very far from the coal feed point; with the large particles, however, the same amount of volatiles is released more slowly and the heat release is spread more uniformly. With the small particles, therefore, the rate of heat release near the coal feed point may be high and, because of the large amount of volatiles released locally, the atmosphere near the coal feed point may be a reducing atmosphere; the metallurgical implications of this have to be considered. When the large coal particles are fed to the bed, the emission of volatiles is more gradual than with small particles, so that the distribution of heat release is more uniform and the likelihood of a severe reducing atmosphere locally is less (Gibbs 1987).

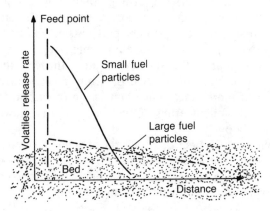

Figure 5.20 The transverse distribution of the rate of release of volatiles from a given coal for two different sizes of coal feed.

5.6.4.3 Summary
It is hoped that the above has provided just a small insight into some

aspects of fluidized bed combustion and the principles involved—it has perforce had to be somewhat selective, concentrating on bubbling fluidized bed systems—but it should provide a basis for developing the further skill and arts needed for successful commercial design. A brief review of other topics follows.

5.7 Distinction Between Boilers and Furnaces

The output of a fluidized bed boiler is steam or hot water, whereas the output of a fluidized bed furnace is hot gases. The former operate with relatively low amounts of excess air for reasons of efficiency. With furnaces, however, the amount of excess air supplied is normally much greater than with a boiler, so as to avoid excessive bed temperatures, and this means that the firing rate (MW m^{-2} of bed plan area) is smaller than a fluidized bed boiler having cooling tubes in the bed.

The design of a furnace is dependent upon the use to be made of the hot gases—in many cases combustion products direct from the fluidized bed furnace are acceptable. If hot air is needed then it may be generated by installing air cooled tubes in the bed and in the exhaust gas duct; however, the design of such tubing requires considerable skill and experience. Nevertheless, such furnaces are commercially available.

5.8 Methods of Starting Up

Firstly, before solid fuel can be fed to a fluidized bed combustor and the subsequent combustion be self-sustaining, the bed temperature has to be raised sufficiently for the fuel to ignite when introduced into the bed. The required temperature is thus dependent upon the fuel ignition temperature. The second requirement is that oxygen has to reach the fuel particles at a sufficient concentration for the fuel particles to burn and release heat sufficiently rapidly for them to maintain combustion.

A fluidized bed combustor contains a large mass of inert solids and, by virtue of in-bed tubing and a continual flow of initially cool fluidizing air, heat removal from the bed is high. So, the first problem in starting up tends to be one of transferring heat into the bed at a rapid rate, to exceed the heat losses.

5.8.1 Methods for initial heating of the bed
The fastest method of heating the bed up to a high temperature, e.g. 850 °C, is to burn fuel gas in it. This requires a properly designed safety and control system. Good engineering and control practice should ensure this. Some guidance on burning gas in fluidized beds can be found in a chapter by Broughton in Howard (1983).

However, with many fuels, e.g. bituminous coals, it is not essential to raise the temperature of the bed to more than about 450 °C for a satisfactory transition to self-sustaining combustion. This can be achieved either by hot gas start-up, in which hot combustion products from a gas burner which contain a lot of excess air are supplied to the plenum chamber of the fluidized bed, or alternatively, by directing a highly radiant burner flame on to the surface of a bed which is gently fluidized. If hot gases are supplied to the plenum chamber a water cooled distributor is generally required, in order to avoid distortion. Alternative methods of start-up, which also include burning oil in the bed, and modelling of start-up have been reported on in some detail by Moodie and Vickers (1985).

5.8.2 Transition from initial heating to self-sustaining combustion

The second requirement for a start-up system is for it to be capable of ensuring a satisfactory transition (i.e. without loss of control of bed temperature), from commencement of solid fuel feeding to self-sustaining combustion. Thus, the initial heat input to the bed can be shut off or gradually reduced as the main fuel starts to deliver heat. This situation is shown schematically in figure 5.21.

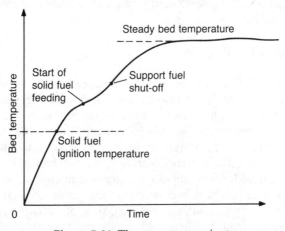

Figure 5.21 The start-up transient.

With large, multi-bed fluidized bed combustors the options are to use one of the beds for start-up and gradually bring the other beds into operation as needed or, alternatively, to provide each bed with its own start-up system.

Whatever the plant, designers have to try to make start-up systems as well as on-load control fully automatic.

5.9 Circulating or 'Fast' Fluidized Bed Combustion Systems

Figure 5.22 shows the essentials of a circulating or 'fast' fluidized bed combustion system. The difference between this type of system and the bubbling bed, in which elutriated solids are recycled back to the bed in order to achieve greater carbon burnout (as depicted in figure 5.12), is that the velocity of the fluidizing gas is significantly higher, typically 5–10 m s^{-1}, than with bubbling beds. Thus, the bed is highly expanded, operating in the fluidization regime between bubbling or turbulent fluidization and pneumatic transport.

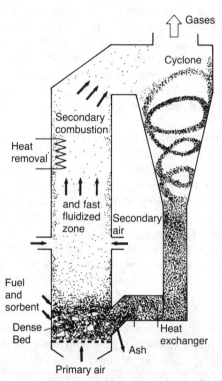

Figure 5.22 The 'fast' or circulating fluidized bed combustion system.

The high air velocity above the secondary air inlets of the circulating fluidized bed combustion system, typically 5–10 m s^{-1}, leads to a small planform area for a given duty, despite a tall containing vessel. However, the small cross section reduces the need for a large number of fuel feed points.

Fuel and sorbent are fed to the bubbling or turbulently fluidized bed supported on the distributor in the lower part of the system. There is a transition, although not sharply defined, in suspension density with height from this regime to the pneumatic transport or fast regime where the concentration of particles in the gas/particle suspension is quite dilute. In order to maintain a constant mass of solids in the system, solids have to be returned to the bed at the same rate as they are elutriated from it.

Operation in the highly expanded bed regime results in improved gas-to-particle contacting, so that high carbon burnout efficiencies and low emissions of oxides of sulphur and nitrogen with high utilization of sulphur sorbent have been claimed. The combustion efficiency is influenced by the efficiency of the high-temperature cyclone which separates the particles from the gas stream, where and how the fuel is fed, and how far the secondary air penetrates (Kullendorff and Andersson 1986). Details of the fast fluidization regime and exploitation of it are described by Reh (1971), Yerushalmi *et al* (1976), Basu (1986) and the various proceedings of international conferences on fluidized bed combustion.

Different variants of the system shown in figure 5.22 are available (see, for example, Engstrom and Sahagian 1986, Nack *et al* 1980, Reh *et al* 1980, Stockdale and Stonebridge 1985, Kullendorff and Andersson 1986) but, whatever the variant, circulating or 'fast' fluidized beds still have to obey the laws of thermodynamics! Heat has still to be removed from the combustion system to keep the temperatures lower than that at which ash softens. The places at which heat may be conveniently removed are through the reactor walls and from the circulating solids, by means of a heat exchanger placed in the recycle loop, as shown in figure 5.22. Different designers have differing opinions as to the preferred place and quantity of heat extraction, subject of course to satisfying the overall heat removal requirements. With coal burning, circulating fluidized bed combustion systems it is common to find that solids are being recycled at many times the coal feed rate; clearly this offers the possibility of a large amount of heat removal in the external heat exchanger if desired. It will be left as an exercise for the reader to estimate the distribution of heat removal in such a case.

5.10 Control of Emissions

5.10.1 Introduction
The emissions of pollutants from industrial plants have become the subject of increasing public concern and legislation, limiting the amount

of pollutants which may be emitted, is in force in industrialized countries. The pollutants in the exhaust gases from combustion systems which have attracted the greatest attention are sulphur dioxide, SO_2, the oxides of nitrogen, commonly referred to as NO_x, and particulates. Normally, the particulates are removed from the exhaust gases satisfactorily by passing them through well designed gas cleaning equipment, e.g. cyclones, electrostatic precipitators, bag filters etc.

The sources of the above two gaseous pollutants from burning fossil fuels are the sulphur and nitrogen contained in the fuel and, although in fluidized bed combustion systems it plays little part on the formation of oxides of nitrogen, atmospheric nitrogen. Fluidized bed combustion systems have the capability of limiting the emission of these gaseous pollutants to amounts acceptable to legislation. It is important for designers to acquaint themselves with the most up-to-date limits on the emission of pollutants imposed by legislation in the country in which the plant to be designed is to be located. To obtain some perspective of the problem and noting that different countries use different criteria to define pollutant concentrations, something of the order of 90% of the sulphur dioxide is likely to have to be captured; also, in future, the concentration of NO_x in exhaust gases may be limited to less than about 250 ppm.

Despite the great amount of successful work carried out to develop satisfactory methods for controlling the emission of pollutants, some of the mechanisms involved are not completely understood. A discussion of this matter is beyond the scope of this book; details may be found in the literature (see, for example, Stantan 1983, Shaw 1983). The following simplified explanations, however, may be sufficient to provide an elementary understanding of what is involved in making fluidized bed combustion equipment meet emission standards and to indicate the present state of the art.

5.10.2 Emissions of sulphur dioxide

All fossil fuels contain sulphur and certainly, for coals, the sulphur content varies widely, typically from about 1 to 10%. Fortunately for Britain, most British coals have a sulphur content from less than 1% to 2%. Combustion of fossil fuels leads to the release of sulphur dioxide, SO_2, and to a lesser extent sulphur trioxide, SO_3. The technique employed for controlling the amount of SO_2 in the exhaust gases from fluidized bed combustion systems is to feed limestone or dolomite particles to the bed, which act as a sorbent. A large fraction of the sulphur is thus captured in solid form so that the sulphated sorbent particles can then be removed from the bed and also be separated from the exhaust gas stream by the normal gas cleaning equipment.

Limestones contain calcium carbonate ($CaCO_3$) and dolomites contain calcium magnesium carbonate ($CaCO_3 \cdot MgCO_3$). Considering first limestone, Stantan (1983) pointed out that in an oxidizing atmosphere, sulphur dioxide can react with calcium carbonate direct to form calcium sulphate ($CaSO_4$):

$$CaCO_3 + SO_2 + \tfrac{1}{2}O_2 \rightleftharpoons CaSO_4 + CO_2 + 303 \text{ MJ kmol}^{-1}. \quad (5.31)$$

Also, if calcium carbonate is heated, e.g. when limestone is added to a hot bed, it becomes 'calcined' to form calcium oxide (CaO) and sulphur dioxide will also react with calcium oxide in an oxidizing atmosphere to form calcium sulphate:

$$CaO + SO_2 + \tfrac{1}{2}O_2 \rightleftharpoons CaSO_4 + 486 \text{ MJ kmol}^{-1} \quad (5.32)$$

the calcination reaction being the endothermic reaction

$$CaCO_3 \rightleftharpoons CaO + CO_2 - 183 \text{ MJ kmol}^{-1}. \quad (5.33)$$

The situation with dolomite is a little more complex, although similar principles apply. This matter will not be pursued here and the reader is referred to Stantan (1983).

The way and rates at which the above reactions proceed depend upon the type of limestone, its preparation for use, its reactivity and physical properties, e.g. porosity, particle size (limestones differ widely in composition and properties and methods for predicting their performance have been developed, e.g. by Fee *et al* 1982), the temperature of the process, bed depth, fluidizing velocity etc.

Among the more important operating parameters are the calcium:sulphur molar ratio and bed temperature and these vary with the sorbent used. As a rough and ready guide with bubbling fluidized beds operating at atmospheric pressure, a bed temperature of about 850 °C and a Ca : S molar ratio of about 3 or more is likely to be required for a sulphur capture of the order of 90%.

The quantity of sorbent required for emission control and, therefore, the solids handling and storage bunker capacity for the limestone is significant, as the following example shows, and may require to be taken into account in the heat balance as well.

Example 5.6. A fluidized bed combustion system having an output of 30 MW at 79% efficiency when using coal of calorific value 25 MJ kg^{-1} with a sulphur content of 3.8% requires a particular limestone to be fed to it at a calcium : sulphur ratio of 3.1, in order to limit emissions of sulphur dioxide sufficiently. The limestone used contains 85% calcium carbonate. Determine the flow rate of limestone required. (Assume the following atomic weights: S = 32, Ca = 40, C = 12, O = 16.)

$$\text{Coal flow rate} = \frac{\text{output}}{\text{efficiency} \times \text{calorific value}}$$

$$= \frac{30}{0.79 \times 25}$$
$$= 1.52 \text{ kg s}^{-1} = 5470 \text{ kg h}^{-1}.$$

$$\text{Molar flow of sulphur} = \frac{5470 \times 0.038}{32}$$
$$= 6.5 \text{ kmol h}^{-1}.$$

For a Ca:S molar ratio of 3.1 the calcium molar flow rate is

$$3.1 \times 6.5 = 20.15 \text{ kmol h}^{-1}.$$

The molecular weight of calcium carbonate is 100, hence the mass flow of $CaCO_3$ required

$$= 20.15 \times 100$$
$$= 2015 \text{ kg h}^{-1}$$

and the mass flow of limestone

$$= \frac{2015}{0.85}$$
$$= 2370 \text{ kg h}^{-1}$$

(i.e. 43% of the coal feed rate).

5.10.3 Emissions of oxides of nitrogen

It has to be admitted that knowledge of the mechanisms involved in controlling emissions of oxides of nitrogen is incomplete and that each case has to be investigated experimentally. The reader is referred to Shaw (1983) and papers in the proceedings of international conferences on fluidized bed combustion for a more detailed discussion and data than can be included here.

However, at present, the commonest method of controlling the amount of oxides of nitrogen in the exhaust gases is by carrying out combustion in two separate stages. In the first stage, i.e. the bed, combustion is carried out with insufficient air for complete combustion, the argument being that much of the NO formed by oxidation of part of the fuel nitrogen is reduced by it being absorbed on the surface of char particles. (Precisely where, how and at what rates these formation and absorption processes occur in a given combustor are not known.) Combustion is then completed by admitting secondary air and the temperature of the gases is maintained sufficiently low, e.g. <1000 °C, for oxidation of atmospheric nitrogen to be insignificant.

The optimum ratio of primary to secondary air, the best location at which to introduce it, bed temperatures etc vary from case to case and have to be determined by experiments with the given combustor. There are many influences upon emission of NO_x in addition to these parameters, for example the type of fuel being burnt, its particle size and the fluidization regime of combustor.

There is also some evidence that optimum conditions for abatement of NO_x do not necessarily coincide with those for maximizing sulphur retention.

5.11 Concluding Remarks

Fluidizing bed combustion systems can now satisfy many differing requirements simultaneously, ranging from economic criteria, through reliability of operation, to low emission of pollutants. It is hoped that this chapter has given in an elementary way some indication of what is involved.

The present status of fluidized bed boilers and furnaces has only been achieved by engineering expertise of high order, persistent experimentation, objective analysis of data and dogged determination. Future improvements will require no less.

Examples

1. Two different types of coal are to be burnt in a fluidized bed combustor.

   The first, coal A, has a calorific value as fired of 27 MJ kg^{-1}; for purposes of calculation, coal A may be considered to be of composition as fired 75% carbon, 5% hydrogen, 8% ash and 12% moisture.

   The second fuel, coal B, is of lower grade than coal A and has a calorific value as fired of 16 MJ kg^{-1}; the composition as fired is 46% carbon, 3% hydrogen, 39% ash and 12% moisture.

   If the air and fuel are supplied at atmospheric pressure and at 25 °C, and the products leave at 850 °C, determine the heat which must be removed from the system per kg of fuel and per MW of fuelling if combustion is regarded as being complete; the excess air in each case is 30% and the specific heat of fuel and ash is taken as 0.9 kJ kg^{-1}.

   What ash handling capacity is required for each fuel for a fuelling rate of 60 MW?

2. A fluidized bed furnace operating at atmospheric pressure is required to produce a stream of hot combustion products at a bed temperature of 825 °C. The mass flow of combustion products is to be 7 kg s^{-1} and heat losses from the system are such that the air : fuel ratio by mass is 25 : 1.

   Estimate the planform area required for the bed if the fluidizing velocity is 1.2 m s^{-1}; take the density of air at 825 °C as 0.32 kg m^{-3}. What is the subsequent firing rate (MW m^{-2})?

3. A fluidized bed boiler of thermal output 6 MW operates with an efficiency of 75% and a firing rate of 1.6 MW m^{-2}. Estimate the planform area of the bed and determine the fluidizing velocity if the air : fuel ratio by mass is 13.5. If conditions are such that 40% of the heat liberated by the fuel has to be removed from the bed, estimate the length of tubing of 75 mm diameter which has to be accommodated within the bed if the bed temperature and tube wall temperature are 875 and 95 °C, respectively, and the bed-to-immersed-surface heat transfer coefficient is taken as $0.35 \text{ kW m}^{-2} \text{ K}^{-1}$. Suggest a suitable bed depth to accommodate this amount of tubing.

If the fuelling rate only is reduced by 35% and the bed temperature is unchanged, estimate how much tube surface area would then be required if the bed temperature is to remain unchanged. How might the necessary change of immersed surface area be accomplished?

4. A fluidized bed combustor pilot plant is used for burning coal whose higher calorific value as fired was 25.5 MJ kg^{-1}. The ash and moisture content of the coal as fired are 12.5% and 11%, respectively. An experiment showed that the amount of carbon monoxide and unburnt hydrocarbons in the exhaust gases are both negligible, but that the carbon content of the dust captured by the gas cleaning equipment was 34%.

During the trial 305 kg of dry coal were burnt and the amount of dust captured was 61.4 kg.

Assuming all the coal ash to be captured by the gas cleaning equipment, estimate the efficiency of carbon burnup and the amount of solids elutriated from the bed other than coal ash and carbon. Take the calorific value of carbon as 32.8 MJ kg^{-1}.

What parameters might be altered, or modifications made, to the design of the plant so as to increase the efficiency of carbon burnup?

5. Highly porous, reactive char is to be burnt in a fluidized bed combustor. What is most likely to limit the burning rate?

The ash has a softening temperature in excess of 1100°C. Discuss the choice of fluidizing velocity, bed temperature, and the merits or otherwise of recycling.

6. A fluidized bed combustor is to be used for burning low-volatile, relatively unreactive solid fuel. Experiments on a small scale showed that combustion could not be sustained unless the bed temperature was higher than 730 °C. Tests with the ash showed that it tended to become sticky at a temperature of 950 °C. Discuss the bed operational constraints or difficulties that this fuel might impose.

If the fuel supplied was of narrow size range, with a mean size of the order of 15 mm, discuss the merits of using sand particles of mean size ~2 mm as the main bed material, as against 0.5 mm.

7. A fast fluidized bed combustor burns solid fuel of ash content 48% and higher calorific value as fired of 16.5 MJ kg^{-1} at the rate of 7.8 kg s^{-1}, the total air : fuel ratio by mass being 6.3. All air and fuel are supplied at a temperature of 25 °C and the temperature of the gases and solids, both entering and leaving the cyclone, is 850 °C. The gas leaving the cyclone is also at a temperature of 850 °C. Taking the mean specific heat of the gases leaving the cyclone as 1.14 kJ kg^{-1} K^{-1} and that of the ash as 0.8 kJ kg^{-1}, estimate the rate of heat removal required from the system if the ash leaving the cyclone is uncooled.

The cross section of the fast fluidized zone is square and the fluidizing velocity is 10 m s^{-1}; assuming the suspension to be dilute and the temperature and density in the fast fluidized section to be uniform at 850 °C and 0.31 kg m^{-3}, respectively, determine the planform area of the zone and its side length.

(a) If all the heat has to be removed through the walls of the fast fluidized zone, estimate the height of the fluidized zone, given a value for the heat transfer coefficient at the wall/suspension interface of the fast fluidized zone as 100 W m^{-2} K^{-1} and the mean wall temperature as 320 °C.

(b) The combustor has to be modified to reduce its height by 50%, thus reducing the wall surface area available for heat transfer. In order to retain the same total heat removal rate from the system, it is proposed to remove heat by cooling the recycled solids to 350 °C. Estimate the rate of solids recycle required, taking their specific heat as 0.9 kJ kg^{-1} K^{-1}; express the answer as a multiple of the solid fuel feed rate.

8. The fast fluidized bed combustor of question 7(b) is to be operated at 60% of the fuelling rate, with the same air:fuel ratio. What would the rate of solids recycle be assuming other parameters to be unchanged?

9. How is the emission of sulphur dioxide in the exhaust gases from a fluidized bed combustor kept at a low level? How would the calcium:sulphur ratio influence the concentration of SO_2 emitted? Discuss briefly the consequences of improving combustor design such that the desired emission level of SO_2 could be achieved with a lower calcium:sulphur ratio.

References

Avedesian M M and Davidson J F 1973 Combustion of carbon particles in a fluidized bed *Trans. Inst. Chem. Eng.* **51** 121–31

Basu P 1976 *PhD Thesis* University of Aston in Birmingham

—— (ed.) 1986 *Circulating Fluidized Bed Technology* (Toronto: Pergamon)

Broughton J 1983 in *Fluidized Beds: Combustion and Applications* ed. J R Howard (London: Applied Science) ch 9
Chakraborty R K and Howard J R 1978 Burning rates and temperatures of carbon particles in a shallow fluidized bed combustor *J. Inst. Fuel* **51** 220–4
—— 1981a Combustion of char in shallow fluidized bed combustors: influence of some design and operating parameters *J. Inst. Energy* March, 48–54
—— 1981b Combustion of single carbon particles in fluidized beds of high-density alumina *J. Inst. Energy* March, 55–8
Chakraborty R K and Vickers M A 1984 in *Proc. 3rd Int. Conf. Fluidized Combustion: is it Achieving its Promise?, London, 1984* (London: Institute of Energy) paper DISC/33
Cooke M J 1984 in *Proc. 3rd Int. Conf. Fluidized Combustion: is it Achieving its Promise?, London, 1984* (London: Institute of Energy) pp RAPP/ENV/V/4-1–6
D'Amore M, Donsi G and Massimilla L 1980 in *Proc. 6th Int. Conf. Fluidized Bed Combustion, Atlanta, Georgia, 1980* vol 2, pp675–84
Donsi G, Massimilla L and Miccio M 1981 *Combust. Flame* **41** 57–69
Ehrlich S 1976 in *Proc. 4th Int. Conf. Fluidized Bed Combustion, 1975* (McLean, Virginia: MITRE Corporation) pp15–20
—— 1984 in *Proc. 3rd Int. Conf. Fluidized Combustion: is it Achieving its Promise?, London, 1984* (London: Institute of Energy) pp KA/1–29
Engstrom F and Sahagian J 1986 in *Circulating Fluidized Bed Technology* ed. P Basu (Toronto: Pergamon) pp309–16
Evans R L and Anastastiou R B 1985 On the performance of pressurized fluidized bed cycles for power generation *Proc. Inst. Mech. Eng.* A **199** 45–51
Fee D C, Myles K M, Wilson W I, Fan L-S, Smith G W, Wong S H, Shearer J A, Lenc J F and Johnson I 1982 Sulphur control in fluidized bed combustors: methodology for predicting the performance of limestone and dolomite sorbents *Argonne National Laboratory Report* ANL/FE-80-10
Fenton K 1977 in *Technical Data on Fuel* 7th edn ed. J W Rose and J R Cooper (Edinburgh: Scottish Academic) pp18–20
Gibbs B M 1987 private communication
Grace J R 1986 in *Gas Fluidization Technology* ed. D Geldart (Chichester: Wiley–Interscience) ch 11
Highley J and Kaye W G 1983 in *Fluidized Beds: Combustion and Applications* ed. J R Howard (London: Applied Science) ch 3
Highley J and Merrick D 1971 The effect of spacing between solid feed points on the performance of a large fluidized bed reactor *AIChE Symp. Ser. No 116* **67** 219–27
Howard J R (ed.) 1983 *Fluidized Beds: Combustion and Applications* (London: Applied Science)
International conferences on fluidized bed combustion
  In the United States:
  3rd, 1973, Environmental Protection Agency EPA-650/2-73-053
  1976 4th, 1975, US Energy Research and Development Administration (Maclean, Virginia: MITRE Corporation)
  5th, 1977, Washington DC
  6th, 1980, Atlanta, Georgia US/DOE/CONF-800428

7th, 1982, Philadelphia US/DOE/METC 83-48
8th, 1985, Houston, Texas US/DOE/85/6021
9th, 1987, Boston, Mass., ASME, New York
In the United Kingdom:
1st, 1975, Institute of Fuel (now the Institute of Energy) Fluidized Combustion, Symp. Ser. No 1
2nd, 1980, London, Institute of Energy, Fluidized Combustion: Systems and Applications, Symp. Ser. No 4
3rd, 1984, London, Institute of Energy, Fluidized Combustion: is it Achieving its Promise?

Kullendorff A and Andersson S 1986 in *Circulating Fluidized Bed Technology* ed. P Basu (Toronto: Pergamon) pp83–96

La Nauze R D 1985 in *Fluidization* 2nd edn, ed. J F Davidson, R Clift and D Harrison (London: Academic) ch 16

Minchener A J, Read D C and Stringer J 1984 in *Proc. 3rd Int. Conf. Fluidized Combustion: is it Achieving its Promise?, London, 1984* (London: Institute of Energy) paper DISC/20

Moodie J and Vickers M A 1985 Some considerations of start up in the design of fluidized beds *Int. J. Energy Res.* **9** 203–9

Nack H, Anson D and DiNovo S T 1980 in *3rd Int. Conf. Fluidized Combustion: Systems and Applications, London, 1980, Inst. Energy Symp. Ser. No 4*, paper VI-4

Peçanha R P and Gibbs B M 1984 in *Proc. 3rd Int. Conf. Fluidized Combustion: is it Achieving its Promise?, London, 1984* (London: Institute of Energy) paper DISC/9

Pentland I W and Mitchell T 1984 in *Proc. 3rd Int. Conf. Fluidized Combustion: is it Achieving its Promise? London, 1984* (London: Institute of Energy) DISC/22/190–8

Pillai K K 1985 Devolatilization and combustion of large particles in a fluidized bed *J. Inst. Energy* March, 3–7

Prins W and van Swaaj W P M 1986 in *Fluidized Bed Combustion* ed. M Radovanovic (New York: Hemisphere) pp169–84

Reh L 1971 Fluidized bed processing *Chem. Eng. Prog.* **67** 53–63

Reh L, Schmidt H W, Daradimos G and Petersen V 1980 in *3rd Int. Conf. Fluidized Combustion: Systems and Applications, London, 1980, Inst. Energy Symp. Ser. No 4*, paper VI-2

Roberts A, Pillai K K and Stantan J E 1983 in *Fluidized Beds: Combustion and Applications* ed. J R Howard (London: Applied Science) ch 4

Rose J W and Cooper J R (ed.) 1977 *Technical Data on Fuel* 7th edn, British National Committee of the World Energy Conference (Edinburgh: Scottish Academic)

Ross I B and Davidson J F 1981 The combustion of carbon particles in a fluidized bed *Trans. Inst. Chem. Eng.* **59** 108–14

Shaw J T 1983 in *Fluidized Beds: Combustion and Applications* ed. J R Howard (London: Applied Science) ch 6

Skinner D G 1970 *The Fluidized Combustion of Coal – A Review of the Fundings of Research up to the Beginning of 1969* (London: National Coal Board)

Squires A M 1983 in *Fluidized Beds: Combustion and Applications* ed. J R Howard (London: Applied Science) ch 8
Squires A M, Kwauk M and Avidan A A 1985 Fluid beds: at last, challenging two entrenched practices *Science* **230** 1329–37
Spalding D B and Cole E H 1973 *Engineering Thermodynamics* 3rd edn (London: Edward Arnold)
Stantan J E 1983 in *Fluidized Beds: Combustion and Applications* ed. J R Howard (London: Applied Science) ch 5
Stockdale W and Stonebridge R 1985 in *Coaltech 85, Int. Conf. Exhib. Coal Utilization and Trade, London, 1985* (International Presentations Group, Schiedam, Netherlands)
Turnbull E and Davidson J F 1984 Fluidized combustion of char and volatiles from coal *AIChE J.* **30** 881–9
Yerushalmi J, Gluckman M J, Dobner S, Graff R A and Squires A M 1976 in *Fluidization Technology* vol II, ed. D L Keairns (Washington, DC: Hemisphere) pp437–69

Bibliography

Beer J M, Massimila L and Sarofim A F 1980 in *3rd Int. Conf. Fluidized Combustion: Systems and Applications, London, 1980, Inst. Energy Symp. Ser. No 4*, paper IV-5
Davidson J F 1980 in *3rd Int. Conf. Fluidized Combustion: Systems and Applications, London, 1980, Inst. Energy Symp. Ser. No 4*, paper RAP-IV
—— 1982 in *The Chemical Industry* ed. D Sharp and T F West (Chichester: Ellis Horwood) ch 9
—— 1984 in *Proc. 3rd Int. Conf. Fluidized Combustion: is it Achieving its Promise?, London, 1984* vol 2 (London: Institute of Energy) paper REV/1
Davidson J F, Clift R and Harrison D (ed.) 1985 *Fluidization* 2nd edn (London: Academic)
Garbett E S and Hedley A B 1980 in *3rd Int. Conf. Fluidized Combustion: Systems and Applications, London, 1980, Inst. Energy Symp. Ser. No 4*, paper IV-6
Hamblin F D 1971 *Abridged Thermodynamic and Thermochemical Tables* (Oxford: Pergamon)
Jung K and La Nauze 1984 in *Fluidization* ed. D Kunii and R Toei (New York: Engineering Foundation) pp427–34
Keairns D L (ed.) 1976 *Fluidization Technology* (Washington, DC: Hemisphere)
Kunii D and Toei R (ed.) 1984 *Fluidization* (New York: Engineering Foundation)
Macdonald D M and Stoton J 1981 in *Fluidized Combustion: Systems and Applications, London, 1980, Inst. Energy Symp. Ser. No 4*, paper 1B-4
Poersch W and Zabeschek G 1980 in *3rd Int. Conf. Fluidized Combustion: Systems and Applications, London, 1980, Inst. Energy Symp. Ser. No 4*, paper IV-2
Radovanovic M (ed.) 1986 *Fluidized Bed Combustion* (New York: Hemisphere)

Ross I B, Patel M S and Davidson J F 1982 The temperature of burning carbon particles in a fluidized bed *Trans. Inst. Chem. Eng.* **60** 83

Sharp D and West T F 1982 *The Chemical Industry* (Chichester: Ellis Horwood)

Van der Post A J, Bosgra O H and Boelens G 1980 *3rd Int. Conf. Fluidized Combustion: Systems and Applications, London, 1980, Inst. Energy Symp. Ser. No* 4, paper IV-3

6 Closure

6.1 Introduction

It will have been seen that processes involving intimate contact between particles and gases can be carried out effectively by fluidized bed techniques. When applied on an industrial scale, the choice of the most appropriate system is governed by economics as well as the technology.

The principles laid down in this book form a vital part of the engineering of fluidized bed types of chemical process reactors, catalytic crackers, boilers, furnaces, incinerators, heat exchangers, calciners, gasifiers, driers, metallurgical heat treatment furnaces, roasters, solids transport systems and dry methods of flue desulphurization, while new applications are continually being experimented with. The importance of good design and operating practice, based on reliable data, careful observation and practical experience can hardly be overstated. Boilers, furnaces and heat exchangers have been discussed in earlier chapters. It may be helpful, however, to conclude the book with the briefest of comments on some of these other applications and give references for further reading. Design and scale-up of plant—a topic of major importance—has had to be treated similarly.

6.2 Gasifiers

Gasifiers are used to produce fuel gas from coal or other feedstocks such as wood, biomass or refuse. Gasification processes involve heterogeneous gas–solid chemical reactions, which require substantial amounts of heat and mass transfer, and the problems posed in the design and development of plant to carry out these processes have led to a variety of types of gasifiers. Schilling *et al* (1981) review fundamental types of gasification plant current at that time, while Lacey (1988) presents the most recent review on the gasification of coal. Industrial implementation of gasifiers depends upon the economics of each application and this is affected considerably by the cost of the feedstocks and the efficiency of conversion.

Fluidized bed gasifiers are one type of gasifier. They have the advantages of uniform temperature distribution throughout the gasifier and the adaptability to varying load demand without serious loss of efficiency. These gasifiers can be used to produce a low sulphur content gas, sufficient to meet environmental requirements, either by addition of limestone or dolomite with the coal feed to absorb the sulphur, or by post-gasification purification techniques. Figure 6.1 from Green (1987) shows the flow sheet for a fluidized bed gasification plant used for gasifying coal in the presence of limestone, while figure 6.2 shows a further development of the oxygen donor gasification process described by Moss (1983), which utilizes fluidized bed technology to produce a nitrogen-free gas without the need for an oxygen plant.

In addition to their use as producers of fuel gas for a variety of industrial processes, the use of coal gasification plant is one route to pollution limited electricity generation incorporating a combined gas turbine and steam plant (see, for example, Schmitt 1981, Lacey 1988). Cleaning the hot product gas, raising the efficiencies by waste heat recovery and processes such as the recycling of fines, oxygen enrichment of the fluidizing gas, improving coal feeding methods etc are under continuous development. Further references include Grainger and Gibson (1981), Green et al (1984) and Butt et al (1987).

6.3 Dryers

Moist substances may be dried by passing hot, dry gases through the damp material. Irrespective of whether or not the material to be dried is fluidizable, a supply of hot, relatively dry gas is needed. In some cases, the product to be dried will tolerate being dried by hot gases from combustion processes, whereas other products are less tolerant and require uncontaminated gas such as clean, hot air. Fluidized bed furnaces have been developed as hot gas generators for crop drying, cement manufacture, etc. Some types of furnace provide hot combustion products while others provide clean hot air (see, for example, Highley and Kaye 1983). The source of the hot dry gas is a separate entity from the dryer itself and of course has to be matched to the dryer.

Reay (1986) gives an account of the fundamentals of drying processes using fluidization techniques and discusses the features of several types of dryer and process design. If the product is capable of being fluidized in the moist state, e.g. granular material, then good particle/gas contacting, relatively gentle handling and lack of moving parts in fluidized bed dryers makes them attractive. Figure 6.3, taken from Reay (1986) shows one type, namely a shallow fluidized bed continuous 'plug flow' flow dryer, a variant of which is the vibrated fluidized bed dryer.

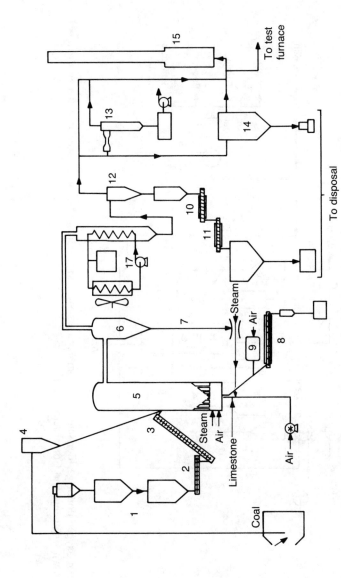

Figure 6.1 A process flow diagram for a gasification pilot plant. 1 Coal feed lock hopper. 2 Metering screw feeder. 3 Transport screw. 4 Bed feed hopper. 5 Gasifier. 6 Primary cyclone. 7 Fines recycle leg. 8 Metering screw feeder. 9 Start-up burner. 10 Fines cooler conveyor. 11 Fines transfer conveyor. 12 Secondary cyclone. 13 Gas scrubber. 14 Bag filter. 15 Gas flare. (Redrawn from Green 1987.) Reproduced by permission of J S Harrison pp British Coal.

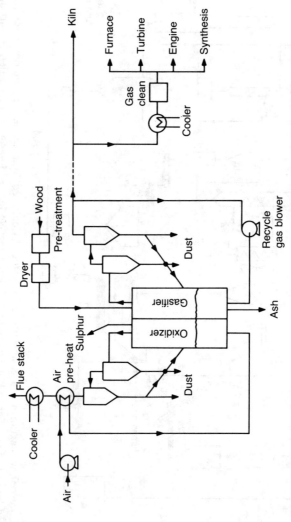

Figure 6.2 An oxygen donor gasifier-schematic-commercial system(s). Redrawn by permission of Wellman Process Engineering Ltd and John Brown Engineers and Constructors Ltd.

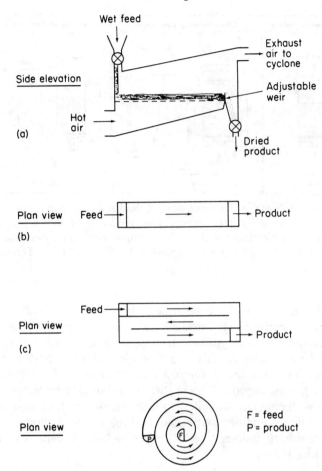

Figure 6.3 A continuous 'plug flow' fluid bed dryer: (a) straight path; (b) reversing path; (c) spiral path. (From Reay 1986.) ©1986. Reprinted by permission of John Wiley & Sons, Ltd.

6.4 Metallurgical Heat Treatment Furnaces

Virr (1983) has described gas fired fluidized bed furnaces used for the heat treatment of metals, notably for hardening and tempering. They have also been used for quenching (see Sommer 1986).

Advances continue to be made, increasing the number of processes to which fluidized bed furnaces are applied and reducing the consumption of the atmosphere gas fluidizing the bed, for example by a fluid-pulse technique (see Sommer 1985) and by recirculation of part of the fluidizing gas (see Fukuda and Hattori 1988). Figure 6.4 shows an example of the latter type of system.

198 Closure

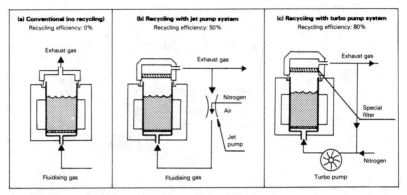

Figure 6.4 A schematic representation of fluidized bed metallurgical furnaces with: (a) no gas recirculation; (b) gas recirculation with a jet pump system; (c) gas recirculation with a turbo pump system. (From Fukuda and Hattori 1988.) Reproduced by permission of Wolfson Heat Treatment Centre.

Fluidized bed furnaces have been developed to include surface treatments including carburizing, nitriding and steam tempering. Conditions for carrying out the boost/diffuse technique used for carburizing in a conventional carburizing furnace can be improved when performed in a fluidized bed heat treatment furnace (see Sommer 1987, Fukuda and Hattori 1988). This improved boost/diffuse carburizing cycle lends itself to accurate mathematical modelling, which enables both process simulation and on-line computer optimized control to be exploited. Other recent references to fluidized bed furnaces include Kubara *et al* (1988) and Mackenzie (1988).

6.5 Solids Transport Systems

Transport of solids, both within fluidized systems and from point to point, are vital to many processes and the reliability of the systems for doing this is paramount. Fluidization based techniques can often offer the best solution to solids transport problems and several types have been developed over the years. Knowlton (1986), to whom the reader is referred for further elucidation of this topic, has given a recent account of the state of the art of transporting solids by fluidized techniques and controlling their flow.

Regrettably, as Knowlton points out, the behaviour of such systems and the characteristics of the several fluidization regimes encountered in the transport of gas–solids mixtures are not fully understood, so that the design of such systems relies heavily upon practical experience with similar problems. In addition to solids conveying systems, such as dense- or dilute-phase pneumatic transport systems, standpipes, etc, non-mechanical valves, which have no moving parts and are cheap, can be used to control or feed solids into equipment such as fluidized beds, or solids conveying systems. Figure 6.5, taken from Knowlton (1986), shows three commonly used types of non-mechanical valves. When aeration gas is supplied to the valve at a sufficient rate, solids flow can commence and the rate is controlled by the flow rate of the aeration gas. Such devices have proved useful and reliable in practice.

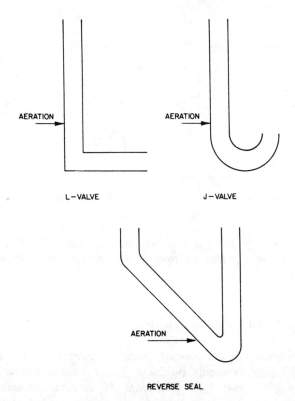

Figure 6.5 The three most common types of non-mechanical valves: L-valve, J-valve, and reverse seal. (From Knowlton 1986.) ©1986. Reprinted by permission of John Wiley & Sons, Ltd.

6.6 Flue Gas Desulphurization

Graf (1986) describes the desulphurizing of flue gases from a power station fired by sulphurous coal. The basic arrangement is as shown in figure 6.6 (from Graf 1986), incorporating the circulating fluidized bed technique. Flue gases, preferably with entrained solid separated from them, enter a venturi reactor. Here they are mixed with fine particles of hydrated lime injected into the reactor which, because of the intimate gas/particle contact, absorb sulphur readily. These particles are carried out of the reactor because the gas velocity is high and are then captured by a dust separator. They are then recirculated back to the venturi reactor many times to give a very long residence time. The system can also be used in semi-dry mode and further development can be expected.

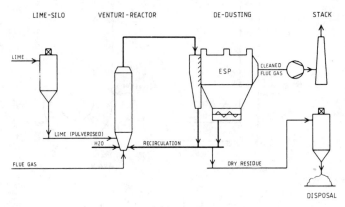

Figure 6.6 Dry flue gas scrubbing using a circulating fluid bed. Reprinted with permission from Graf, Copyright 1986, Pergamon Press PLC.

6.7 Fluidized Bed Catalytic Cracking

Catalytic cracking is the process used for converting heavy oils into gasoline and other valuable products. The reader is referred to Gary and Handwerk (1984) for an account of modern catalytic cracking processes in petroleum refining and also to Venuto and Habib (1979). Fluidized bed catalytic cracking units (sometime abbreviated to FCC units) are employed almost everywhere. There are a number of different configurations and detailed refinements among such plants, but the essence of the process is as follows (see figure 6.7).

(i) Hot, fine ($\simeq 70\ \mu$m) catalyst particles at temperatures in the region of 650–750 °C are injected into the hot oil feed to a fluidized bed reactor at the base of a riser feeding to the reactor. Vaporization occurs and the mixture of hot vapour and entrained catalyst is lifted to the reactor, reacting endothermically as it goes. Most of the catalytic reaction occurs in the riser during transport. The reactor serves mainly to separate the catalyst from the hydrocarbon vapour.

(ii) After disengagement from the catalyst the hot, cracked hydrocarbon vapours are transported for further processing. The separated catalyst particles are heavily coked and are transported to the regenerator via a steam stripping zone.

(iii) The coked catalyst passes to the regenerator whose main purpose is to oxidize the coke on the catalyst and re-expose active catalyst surface ready for reuse in the catalytic cracking reaction. Oxidation of the coke on the surface liberates heat, raising the temperature of the catalyst particles ready for reinjection into the hot oil feed to the reactor.

(iv) Combustion products and a certain amount of entrained catalyst are conveyed upward from the surface of the dense-phase fluidized bed in the regenerator. They have to be captured by cyclones inside the regenerator and return to the bed. The cleaned combustion products then leave the regenerator for further cleaning and heat recovery from the hot combustion products.

(v) The reactor and regenerator constitute two fluidized systems which exchange solids continuously by fluidized transport systems. Large amounts of heat and mass are therefore being transferred.

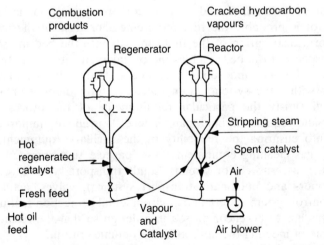

Figure 6.7 The essentials of a catalytic cracking plant. (Adapted from Gary and Handwerk 1984.)

It should be noted that fluidized bed catalytic cracking units exploit the use of fine particles, whose fluidization characteristics correspond to

Geldart's Group A particles (see Chapter 2). A whole body of fluidized bed technology based upon fine particle technology has thus developed, but is beyond the intended scope here. It will be clear that the reliability and control of the solids transport is vital to large-scale catalytic cracking processes and the reader is referred again to Knowlton (1986).

6.8 Design Practice and Scale-up

6.8.1 Design practice

If fluidized bed technology is to be applied successfully, the fundamentals (e.g. basic chemistry, heat and mass transfer and rate-controlling mechanisms) of the processes to which it is being applied have to be clearly understood and so have the practical problems associated with the processes. Indeed, fluidized bed techniques are often proposed so as to deal with these practical difficulties: e.g. fluidized bed combustion processes can be carried out at temperatures below the ash fusion temperature, which inhibits molten ash deposition on heat transfer surfaces, while sulphur dioxide emission can be kept within anti-pollution limits by addition of sorbent to the bed.

As with any other branch of technology, successful exploitation requires that the technology be not only timely, but that it should also incorporate good design and operating practice.

So far as a fluidized bed itself is concerned, a good design of the distributor is probably the first requirement because of the importance of uniform distribution of the fluidizing gas into the bed in most cases. Other aspects of the bed, such as the containment, its lining, the ducting and ports for gas and solids inlet and outlet, solids circulation, and transfer within the system, often require careful thought in order that they will satisfy the particular requirements of the process, operate reliably and require minimal maintenance. It is equally important to put effort into ensuring the reliability of the ancillary equipment, such as feeders, gas cleaning equipment, fans, particle size preparation equipment such as crushers or screens, solids transport systems (e.g. risers, downcomers and pneumatic transport systems), valves, actuators, the plant control system, safety and environmental protection features.

All this has to be encompassed in a design so that the resultant plant is capable of meeting the appropriate economic criteria.

Reactor modelling is an important aid in the development of techniques, but its predictions need to be used with caution and verified by tests.

6.8.2 Scale-up of plant

Scale-up, namely the design of a large plant or device using predictions based upon the results of tests on smaller-size equipment or predictions from mathematical models, can still be an uncertain art, so that performance of the scaled-up plant may differ significantly from that expected. Thus, invariably, risks arise when scaling up. The possibility of scale effects such as those shown by de Groot (1967), who investigated the amount of bed expansion arising with different sizes of reactors and particle size range (see §2.5.3 and figure 2.10), should be borne in mind when designing plant of a different scale or operating circumstances than that for which data are available.

Scaling down can also lead to difficulties; for example, fitting items of a standard size into, say, the passageways of a smaller plant, thus restricting the flow, increasing the pressure drop, or causing consequent problems. However, it may be worth while bringing together here some of the well known aspects and hazards encountered when scaling up.

(i) Performance data obtained with tests with particles belonging to one category are not necessarily applicable to particles of a different category.

(ii) Tests with small-scale equipment may not be possible at the fluidizing velocities envisaged for large-scale plant, e.g. the wall effect and slugging regime encountered with bench-scale apparatus at relatively modest fluidizing velocities.

(iii) The performance of small-scale equipment may be strongly influenced or in some cases dominated by wall effects because the thickness of the boundary layer may occupy a very significant fraction of the vessel diameter. However, with large-scale equipment the boundary layer occupies a much smaller fraction, so that wall effects are confined to a smaller fraction of the cross sectional area.

(iv) If the gas–solid reactions are such that the reactor performance, e.g. conversion obtained, is influenced by the fluid mechanics of the reactor, such as the transfer of gas between the bubble and particulate phases, then considerable care has to be be exercised in modelling or in using baffles to control bubble growth (see Grace 1986) and in the interpretation of small-scale tests.

(v) One of the time-honoured methods used to estimate the size of a reactor to cope with a change in scale of the operation is to maintain the bed depth constant but size the diameter or cross section to accommodate the gas flow (Grace 1974). Generally, in view of the importance of the distributor, the two distributors should preferably be of identical scale, rather than scaled up or down.

(vi) Two-dimensional beds, while being useful for obtaining visual indications, are unlikely to exhibit the same bubble size distribution and coalescence pattern as those with a three-dimensional bed.

(vii) Bed behaviour can change significantly with temperature and pressure, so that the behaviour and data from tests conducted at atmospheric pressure and temperature are not necessarily reproducible at elevated temperature and/or pressure.

6.9 Final Comments

Throughout its history, fluidized bed technology has developed largely from experimentation. The modelling of fluidized bed processes has to reflect this and, because of the complexity of real processes, caution has to be exercised in applying it. The design of fluidized bed plant is likely to continue to be something of an art and rely considerably upon observed behaviour and data obtained from previously constructed plant or pilot plant. However, underlying principles have evolved gradually through experience. It is hoped that this book has provided sufficient instruction on such principles, sufficient examples of applications, and sufficient references to allow the reader, especially a beginner or student, to pursue the subject further with confidence!

References

Basu P (ed.) 1986 *Circulating Fluidized Bed Technology* (Toronto: Pergamon)
Butt A R, Bower C J, Green R C and Patterson N 1987 Coal fired appliances for process heating *Inst. Chem. Eng. Symp. Ser. No 105*
Fukuda T and Hattori H 1988 Surface heat treatments using fluidized beds *Heat Treat. Met.* **15** 53–8
Gary J H and Handwerk G F 1984 *Petroleum Refining – Technology and Economics* 2nd edn (New York: Marcel Dekker)
Geldart D (ed.) 1986 *Gas Fluidization* (New York: Wiley)
Grace J R 1974 Fluidization and its application to coal treatment and allied processes *AIChE Symp. Ser. No 141* **70**
—— 1986 in *Gas Fluidization* ed. D Geldart (New York: Wiley) ch 11
Graf R 1986 in *Circulating Fluidized Bed Technology* ed. P Basu (Toronto: Pergamon)
Grainger L and Gibson J 1981 *Coal Utilisation: Technology, Economics and Policy* (London: Graham & Trotman)
Green R C 1987 Fluidised bed coal gasification for industrial application *Paper presented at Conf. ME187 – Maitrise de l'Energie dans l'Industrie, Paris, 1987*
Green R C, Patterson N P and Summerfield I R 1984 Demonstration of fluidised bed gasification for industrial applications *Energy World* No 116, 7–10
Highley J and Kaye W G 1983 in *Fluidized Beds: Combustion and Applications* ed. J R Howard (London: Applied Science) ch 3
Howard J R (ed.) 1983 *Fluidized Beds: Combustion and Applications* (London: Applied Science)

Knowlton T M 1986 in *Gas Fluidization* ed. D Geldart (Chichester: Wiley–Interscience) ch 12

Kubara M, Jeziorski L and Jasinski J 1988. The kinetics of carburising in electrothermal fluidized beds *Heat Treat. Met.* **15** 59–63

Lacey J A 1988 Gasification: a key to the clean use of coal – parts 1, 2 *Energy World* 155, 156, February, March

Mackenzie R T 1988 The role of 'TACT' fluidized bed heat treatment in just-in-time manufacturing *Heat Treat. Met.* **15** 64–7

Moss G 1983 in *Fluidized Beds: Combustion and Applications* ed. J R Howard (London: Applied Science) ch 7

Reay D 1986 in *Gas Fluidization* ed. D Geldart (Chichester: Wiley–Interscience) ch 10

Schilling H-D, Bonn B and Kraus U 1979 *Coal Gasification* (Essen: Glückhauf GmbH) (Engl. transl. 1981 publ. Graham & Trotman)

Schmitt R W 1981 The 1980 Robens Coal Science Lecture: Coal based electricity in the United States *J. Inst. Energy* June, 63–75

Sommer P 1985 Fluid-pulse heat treatments in fluidised bed furnaces *Heat Treat. Met.* **4** 99–102

—— 1986 Quenching in fluidised beds *Heat Treat. Met.* **2** 39–44

—— 1987 Carburising in fluidised beds *Heat Treat. Met.* **1** 7–10

Venuto P B and Habib E T 1979 *Fluid Catalytic Cracking with Zeolite Catalysts* (New York: Marcel Dekker)

Virr M J 1983 in *Fluidized Beds: Combustion and Applications* ed. J R Howard (London: Applied Science) ch 10

Appendix A

Standard texts on heat transfer show that transient cooling of a sphere in a stream of cooling fluid, from an initial temperature T_i to a temperature T_r at a time t after immersion in a cooling fluid, whose temperature is T_f, may be described by the relationship between dimensionless groups of variables:

$$\frac{T_i - T_r}{T_i - T_f} = \text{Function } (Fo, Bi, (r/R)) \tag{A.1}$$

where Bi is the Biot number (hL/k_s), with $L =$ (diameter/6)
Fo is the Fourier number $(\alpha_s t/L^2)$, with $L =$ (diameter/6)
k_s is the thermal conductivity of the solid
α_s is the thermal diffusivity of the solid $(k_s/\rho_s C_{ps})$
R is the outer radius.

Equation (A.1) thus accounts for the internal thermal resistance of the sphere to heat flow as well as that due to the thermal resistance of the gas film at the outer surface; the relationship given by equation (A.1) is available in the form of charts, which should normally be used to evaluate relaxation times. However, for small particles, such as those in this case, if the internal resistance to heat flow (characterized by the Biot number) is regarded as being very small, a simpler form equation of equation (A.1), namely equation (A.2), may be used to predict 'relaxation time'. Equation (A.2) will yield a slightly lower value than that predicted from the charts arising from (A.1); however, equation (A.2) should predict a value of sufficient accuracy for the use which will be made of it here, namely to help decide a minimum residence time.

The particle temperature T after time t is then given by

$$\frac{T - T_f}{T_i - T_f} = \exp(-t/\tau) \tag{A.2}$$

where τ is a time constant

$$= \left(\frac{\rho_p d_p C_{pp}}{6h}\right) \tag{A.3}$$

and h is a gas-to-particle heat transfer coefficient.

A suitable value has to be chosen for the gas-to-particle heat transfer coefficient. Accordingly, use that predicted by equation (3.4), namely

$$Nu_{gp} = 0.03\, Re^{1.3}. \tag{3.4}$$

The Reynolds number should at this stage be based upon a modest fluidizing velocity, say $0.5\,\mathrm{m\,s^{-1}}$ at a bed temperature of 45 °C, so as to yield a larger relaxation time.

Using the data in table 4.4 yields the Nusselt number:

$$Nu_{gp} = 0.03 \times [(1.104 \times 0.5 \times 0.52 \times 10^{-3})/(1.930 \times 10^{-5})]^{1.3}$$

$$= 1.00$$

giving a gas-to-particle heat transfer coefficient, h_{gp}, of

$$h_{gp} = 1.00 \times (2.762 \times 10^{-5})/(0.52 \times 10^{-3})$$

$$= 0.0531\,\mathrm{kW\,m^{-2}\,K^{-1}}$$

i.e. $53.1\,\mathrm{W\,m^{-2}\,K^{-1}}$.

From equation (A.3), the time constant, τ, for the particle is, therefore

$$= \frac{2640 \times (0.52 \times 10^{-3}) \times 0.8}{6 \times 0.0531}$$

$$= 3.45\,\mathrm{s}.$$

For a particle to be cooled through 90% of its ultimate temperature drop, $0.9 \times (200 - 45)$ K, the left-hand side of equation (A.2) has to be 0.1. Thus

$$\exp(-t/\tau) = 0.1$$

giving

$$t = 2.303\tau$$

and, hence, a relaxation time for the particle

$$= 2.303 \times 3.45$$

$$= 7.94\,\mathrm{s}.$$

Index

Archimedes Number
  definition, 31
Arrhenius form, 171
Attrition, 52

Bed-to-surface heat transfer
  estimation of heat transfer
    coefficients, 84–9, 93
  heat transfer coefficient/particle
    size data – shallow and deep
    beds – bare and finned
    tubes, 99
  influence of fluidizing velocity,
    79, 80
  influence of Archimedes and
    Reynolds Numbers, 80
  influence of bed expansion, 100
  influence of bed temperature, 80
  influence of bubble adjacent to
    wall, 83
  influence of particle size and
    category, 80–2
  interphase gas convective
    component, 77, 80
  models, 79
  radiative component, 77
  overall heat transfer
    coefficient, 121
  particle convective component, 77
  particle convective component –
    effect of particle residence
    time, gas properties, particle
    size and specific heat, 78
  promotion of, 78
  radiative component, 81, 82
  Zabrodsky's correlation for
    maximum heat transfer
    coefficient, 80

Bubble phase of bed, 10
Bubbles, 9, 38
  behaviour with various categories
    of particles, 42
  effect of bubble size on bed
    expansion, 43, 44
  coalescence, 39, 45, 46
  fast, 41
  growth with Geldart's category A
    and B particles, 46
  hold-up of bed – de Groot's
    observations, 45
  rise velocity equation for bubbling
    bed, 40
  rise velocity equation for single
    bubble, 39
  slow, 41
  slugging flow, 46
Burning of carbon/char particles
  burning rate, definition, 168
  burning rate, resistances to, 170
  Burning rate – equations, 170–2
  in bed and freeboard, 168
  mechanisms of, 169
  specific burning rate, 168
  within pores, 168

Circulating fluidized bed, 11
Combustion systems
  calorific value of fuel, 147
  efficiency reducers, 145
  enthalpy definition, 146
  enthalpy of reactants and products,
    146
  enthalpy of reaction, 147
  enthalpy/temperature diagram, 147
  first law of thermodynamics, 144
  heat removal estimation, 148, 149

Combustion systems (*cont.*)
  inputs and outputs, 143
  pollutants, 145
  products of combustion – solid fuels, 144, 145
  reactants, 145
  steady flow energy equation, 146
Contact between fluid and solids
  agitation by rotating drum, 3
  falling cloud, 3
  spouted bed, 3
Correlations
  empirical for minimum fluidizing velocity, 32–4
  empirical gas-to-particle heat transfer coefficient, 73

Distributor
  avoidance of flowback of particles, 118, 119
  estimation of number and size of holes, 117, 118
  pressure drop across, 111, 117
  types of, 90, 119
Drag coefficient, 50

Economic criteria
  capital cost, 137
  cash flow, 137–41
  internal rate of return, 137–41
  payback period, 137–41
  running costs, 137
  sensitivity of cost calculations to input data, 141
Elutriation
  consequences of, 52
  due to bubble behaviour, 53
  influence of turbulence in freeboard zone, 53
  rates of, 54
  simple mechanism of, 52
Emissions control
  calcium: sulphur ratio, 184
  chemical reactions, 184
  legislation, 183
  oxides of nitrogen, 182–6
  oxides of sulphur, 182–6
  particulates, 183
  sorbents, 183, 184

Entrainment, 10
Ergun equation, 30, 33, 38
Excess air – effect on losses and efficiency, 164
Extended surfaces, *see* Finned tubes

Fines
  elutriation of, 52
  production of, 52
Finned tubes
  criteria for spacing of fins, 95, 98
  effective surface area, 95
  estimation of effective surface area, 96
  fin efficiency, 96
  length required for prescribed duty in a fluidized bed, 95, 97, 98
  in shallow and deep beds, 98, 99
  influence of bubble size on heat transfer, 99
  influence of location and bed depth on heat transfer, 98
  overall heat transfer coefficient, 97
  overall thermal resistance, 97
Flow patterns of particles and gas near a tube immersed in a fluidized bed, 93
Fluid
  function of in fluidized beds, 1
  influence of flow regime on fluid–particle interaction, 2
Fluidization
  fast, 11, 12
  incipient, 5, 8, 9, 10, 39
  regimes, 12, 13, 38
  two-phase theory of, 38–45
Fluidized bed
  behaviour, 8, 9, 10
  components of, 5
  containment, 5
  distributor, 5
  expansion, 9, 10, 42, 43, 44
  expansion – de Groot's observations, 44, 45
  gas fluidized, 9, 10
  liquid fluidized, 9
  modelling flow through, 27–30
  pressure drop, 27

Index   211

Fluidized bed boilers and furnaces
  design parameters, 155
  firing rates, 155
  fluidizing velocities, 155
  starting up methods, 179, 180
  turn down, 155
Fluidized bed combustion
  carbon concentration – effect of lateral mixing, 175
  circulating fluidized beds, 181, 182
  combined cycle plant, 153, 154
  combustion efficiency, 164
  combustion of fuel particle – sequence of events, 166, 167
  combustion of solid fuels, 150–2
  critical carbon concentration in the bed, 173, 174
  efficiency – definition, 163
  efficiency – contributing factors, 163–4
  elutriation of unburnt carbon, 167
  emission of volatiles, 167
  fast fluidized beds, 181, 182
  feed point spacing, 175
  historical, 143
  nitrogen oxides emission control, 152, 182–6
  pollutant emission control, 152
  pressurized fluidized bed combustion, 153, 154
  size of combustion chamber, 152, 154
  sulphur absorption, 152, 182
  temperature of combustion, 151
  thermal stability of the bed, 173, 174
  two-stage combustion, 185
  volatiles, 175
Fluidized bed combustors
  bed expansion, exploitation of, 161
  bed temperature, 162
  combustion efficiency, 160, 165
  combustion of volatiles, 162
  elutriation, 160
  estimation of size, 156
  fluidizing velocity, 160
  fuel feed points, 156
  grit refiring, 165
  heat release zones, 162

Fluidized bed combustors (cont.)
  heat removal requirements and methods, 156–9
  inert particles – function and influence on design, 159
  limitations of calculations, 159
  parameters, 160
  turn down, 155, 160, 161
  unburnt carbon, 165
Fluidized bed plant
  ancillary equipment, 202
  applications, examples of, 13
  catalytic cracking, 200, 201
  components, 107
  design considerations and criteria, 107–9, 202
  diagram of simple plant, 108
  dryers, 197
  economic criteria, see Economic criteria
  elutriation limiting velocity, 114
  estimation of bed dimensions, fluidizing velocity and bed expansion, 109–14
  examples of types of process, 1
  flue gas desulphurization, 199, 200
  fluidizing velocity limits, 113
  gas residence time, 113
  gasifiers, 192–6
  influence of changing design criteria on reactor size, 114
  metallurgical heat treatment furnaces, 197, 198
  modelling, 204
  non-mechanical valves, 199
  optimum size of reactor, 137–141
  peripheral items, 107
  pilot plant, 108
  pressure drop across bed, 111
  pressure drop across distributor, 112
  pumping power requirement, 111–12
  reactor, 107
  scale-up, 193–5, 202, 203
  solids transport, 198, 199
  specifications for commercial plants, 107

Index

Fluidizing gas
  exchange between particulate and bubble phases of bed, 45
  flow in form of bubbles, 39
  flow patterns within and around bubbles, 41
  flow through particulate phase, 39
  penetration distance required to achieve particle temperature, 74–6
  residence time, 113
  velocity profiles in above-bed zone, 54
Fluidizing velocity
  definition, 8
  empirical correlations for minimum fluidizing velocity, 32–5
  excess above minimum, 40
  experimental determination of minimum fluidizing velocity, 31, 35, 36
  influence on bed behaviour, 10
  minimum, 8, 10, 29, 30, 40
  theoretical estimation of minimum fluidizing velocity, 30–5
  versus bed pressure drop data, 36

Gas–solid contacting
  interaction according to bubble size, 41
  methods – advantages and disadvantages, 6, 7

Heat removal from fluidized beds
  arrangement of in-bed tubing, 122–3
  bed depth and planform area, 126
  estimation of surface area required, 120–2
  influence of distance between in-bed tubes on particle convection, 123
  sand cooler example, 123–7, 134–8
  staging of fluidized beds, 128–33
Heat transfer
  bed to distributor, containing walls and immersed components, 89, 91

Heat transfer (*cont.*)
  bed-to-immersed surface – components of heat transfer, 77–8, 92–4
  beds of particles, 73
  conduction, convection, 70
  dead zones, 90
  gas-to-particle, 72, 73, 74
  gas-to-particle – correlations for heat transfer coefficient in fluidized beds, 76
  influence of bed expansion, 100
  liquid-to-tube surface, 94
  particle-to-wall, 71
  radiative, 72, 81, 101
  to surfaces located above free surface of bed, e.g. splash zone, transport disengagement zone, 100

Interstices, flow through, 2

Logarithmic mean temperature difference, 75

Mixing of particles, 9

Nusselt Number, 73, 74

Packed bed
  difficulties with, 5
  fixed type, 4
  flow through, 26
  flowing type, 4
Particles
  bulk density of a bed of, 23
  categories of – A and C categories, 25
  categories of – B and D categories, 26
  characterization, 15
  classification according to fluidization characteristics, 24, 25
  correction factor for terminal velocity of non-spherical types, 51

Particles (*cont.*)
    cumulative plot of mass fractions of different sizes, 22
    density of individual particle, 10, 23, 24–6, 32
    disengagement height of bed, 8
    effect of bubble path spacing on mixing, 49
    effect of density differences on mixing, 49
    elutriation, 47
    entrainment in gas, 12
    falling in agglomerates, clusters, in isolation, 51
    finer particles – influence on voidage, 22
    flow at containing wall, 91
    flow of when fluidized, 55
    fluidizable, 10
    function of in fluidized beds, 1
    Geldart's classification of, 24, 25
    launch velocity of, 53
    mass fraction of given size, 17, 20
    mean size, 17–19
    mean size – effect of fraction of fines thereon, 45, 46
    mechanism of mixing, 47, 48
    mixing of, 13
    properties, 15, 16
    recycling of, 12, 13
    relative size range, 21
    relaxation time, 125, 206, 207
    residence time in bed, 125
    segregation of, 49
    separation from gas, 12
    shape, 15
    shape factor, 18
    sieve analysis of, 19
    sieve aperture, 19
    size, 15
    size distribution, 18
    size range, 18, 20, 21
    sphericity, 15
    sphericity, experimental estimation of, 37, 38
    surface area of, 17, 22, 23
    surface: volume ratio, 18
    terminal velocity of, 50

Particles (*cont.*)
    thermal time constant, 206, 207
    transportation by bubble action, 48
    volume of, 17
Particulate phase of bed, 10
Pressure drop
    across bed versus fluidizing velocity, 37
    across fluidized beds, 10, 11
    across packed beds, 10, 11, 29, 30
    equations, 28, 29, 30
    Hagen–Poiseuille equation, 27, 28
    versus fluidizing velocity, 10, 11

Residence time
    of bubble passing through bed, 43
    of particles in bed, 12
Reynolds Number
    definition, 29, 30, 73, 74
    related to flow regime through voids, 30

Schmidt Number, 170
Sherwood Number, 169, 170
Slugging
    behaviour, 45, 47
    limitation on small-scale experiments, 46
    mechanism, 46
    pressure drop across bed, 47
    regime, 11
    slugs – dense phase, 47
Solids loading
    in freeboard zone, 53
    in off-gas, 54
Solids transport
    choking, 59
    concentration of gas and solid in, 57
    continuity equation, 57
    dense-phase, 58
    dilute-phase, 55, 58
    hydraulic, 55
    pneumatic, 55
    pressure drops, 58–60
    pressure gradient versus velocity diagrams, 61, 62
    pumping power, 58
    saltation, 59–62

Solids transport (*cont.*)
  slug flow, 56
  under hydrostatic head, 56
  vertical, 62
Spouted bed, 4

Temperature distribution
  particles and gas near
    distributor, 75
Transport
  disengagement height, 115, 116
  disengagement zone, 53
  pneumatic, 13

Voidage
  at minimum fluidizing velocity, 11, 29–31

Voidage (*cont.*)
  definition, 11, 24
  estimation of, 32
  experimental determination of, 23
  of gas–particle suspension, 12
Volatiles
  devolatilization time, 178
  effect of fuel particle size on
    release rate, 177, 178
  influence on combustor design, 175–7
  release rate, 178
  release rate, transverse distribution of, 178
  secondary air provision, 177

Zabrodsky correlation, 80